Lecture Notes in Statistics

Volume 214

Edited by

P. Bickel
P. Diggle
S. E. Fienberg
U. Gather
I. Olkin
S. Zeger

W0234400

More information about this series at
http://www.springer.com/series/694

Marat Ibragimov • Rustam Ibragimov •
Johan Walden

Heavy-Tailed Distributions and Robustness in Economics and Finance

 Springer

Marat Ibragimov
Institute of Economics and Finance
Kazan Federal University
Kazan, Russia

Rustam Ibragimov
Imperial College Business School
London, United Kingdom

Johan Walden
University of California at Berkeley
    Walter A. Haas School of Business
Berkeley, CA, USA

ISSN 0930-0325   ISSN 2197-7186   (electronic)
Lecture Notes in Statistics
ISBN 978-3-319-16876-0   ISBN 978-3-319-16877-7   (eBook)
DOI 10.1007/978-3-319-16877-7

Library of Congress Control Number: 2015941320

Springer Cham Heidelberg New York Dordrecht London

Printed on acid-free paper

Springer International Publishing AG Switzerland is part of Springer Science+Business Media (www.springer.com)

(To Saniya)
Саниягә
M. I.
(To my parents)
әти-әниемә
R. I.
To Nella, Tintin, Felix, and Theo
J. W.

Foreword

For a long time, the presence of (heavy-tailed) power laws, also referred to as Pareto distributions, has been observed in data covering all fields of science and applications. What has been much less studied is the important question: "What are the economic consequences of this observation?" Based on several of the authors' publications, it is precisely this question which is addressed in this interesting book. Heavy-tailed models typically induce a kind of regime switching (non-robust) behavior as a function of the tail-decay parameter; this leads to a fundamental rethinking of important questions like portfolio diversification or the (re)insurance of catastrophic risks. Extreme heavy-tailed behavior (possibly infinite-mean models) should imply risk management caution on behalf of the end-user, decision maker. The authors carefully explain where the pitfalls are in this economically counterintuitive landscape, how best they can be avoided, but also how to optimally structure products and markets in such an environment. Potential applications go well beyond finance, economics, and insurance. Not only financial crises and crashes have catapulted "thinking about extremes, the worst that can happen" to the forefront of the political and regulatory agendas, but also discussions about global warming and the occurrence of natural disasters call for similar methodology.

This book adds economic thinking to statistical modeling, and as such is most highly welcome!

RiskLab, ETH Zurich   Paul Embrechts
Zurich, Switzerland

Preface

The purpose of this book is to provide a fairly detailed introduction to the analysis and applications of heavy-tailed distributions in a number of important problems in economics, finance, risk management, and insurance. The target audience for the book is graduate students in economics, finance, risk management, probability, statistics, and insurance, although the book may also be of value for advanced undergraduate students who have completed a graduate course in probability. It should also be useful for professionals in the financial and insurance industries, risk managers, and for regulators and policy makers concerned with modeling the effects of crises, large fluctuations and extreme values of financial returns, foreign exchange rates, and other key economic and financial indicators and variables. More broadly, we hope that anyone who is interested in a self-contained treatment of the topic at a post graduate level will find this book useful.

The book is based on several published articles, but much effort has been invested into making it self-contained. Specifically, notation has been made consistent across chapters, and we have also adjusted the language, making it less technical than in its original form to make the results easier to digest. The cost of these modifications is a slight loss in rigor. Especially, most proofs have been excluded. The interested reader is referred to the original publications.

The book consists of three chapters. The first chapter provides a brief introduction to heavy-tailed distributions, and their presence and applications within finance, insurance, and economics. The chapter begins with a literature review, followed by a definition of what it means for a risk to have a heavy-tailed distribution. It then discusses the empirical evidence for the presence of heavy-tailed risk distributions in practice in the aforementioned fields. Finally, the chapter discusses the main point of this book, namely that there are limitations to diversification with such risk distributions. Specifically, whereas diversification is preferred by risk-averse agents when risks are thin-tailed (the traditional case that has been extensively studied), it may actually be hurtful for agents to diversify when risks are heavy-tailed (the nontraditional case that this book focuses on). Two examples of heavy-tailed distributions are discussed, namely Cauchy and Lévy distributed risks, to show the limits of diversification in the heavy-tailed case.

Chapter 2 focuses on the theory of diversification with heavy-tailed risks, and the implications for economics, finance, and insurance. The chapter first introduces the important concept of majorization, which allows for a general analysis of portfolio diversification. Specifically, several classes of risk distributions are introduced and analyzed with respect to whether diversification decreases risk (as in the traditional setting) or increases it (as in the nontraditional setting). The chapter then studies the implications for econometric and statistical inference. Finally, it introduces several models to analyze the implications of the results.

First, a model of a reinsurance market for catastrophe insurance is analyzed. The distributions of aggregate losses due to catastrophic events are known to be very heavy-tailed. It is shown in the model that this may explain why it has been challenging to develop well-functioning markets for risk-sharing of catastrophic risks. Specifically, in such markets, a coordination problem may exist where many entities need to agree to sell insurance policies against catastrophic events for a market to start functioning. In the outcome where no insurance is sold, there is a so-called nondiversification trap. We show that nondiversification traps may arise when risk distributions have heavy left tails and insurance providers have limited liability. When they are present, there may be a coordination role for a centralized agency, e.g., government or a regulatory authority, to ensure that risk sharing takes place.

We next introduce a model of financial intermediaries, in which so-called diversification disasters can occur. Specifically, if there are negative externalities to society if multiple financial intermediaries default on their obligations at the same time, then risk-sharing, i.e., diversification, among these intermediaries may be suboptimal. We suggest that historical legislation, e.g., in form of the Glass–Steagall act, may have had a role in avoiding such outcomes.

As a third example, we study the problem of optimal bundling for a multiproduct monopolist providing goods in auctions or for profit-maximizing prices to consumers with heavy-tailed private valuations for these goods. We show that several results in the literature that hold in the traditional setting under thin-tailed valuations are reversed under heavy-tailed valuations. Finally, we considered growth models for firms investing into information about their markets, again showing that several standard results in the literature are reversed when distributions of variables entering their assumptions are heavy-tailed.

The main conclusion of the results in Chap. 2 is that the presence of heavy tails can either re-enforce or reserve the properties of many important models in economics, finance, risk management, insurance, econometrics, and statistics, depending on the degree of heavy-tailedness. This further emphasizes the importance of having robust econometrically and statistically justified inference methods under heavy-tailedness.

Chapter 3 deals with robust inference methods under heavy-tailedness. The chapter discusses widely used approaches to inference on the degree of heavy-tailedness and their main asymptotic properties. The asymptotic analysis provides the key to developing econometrically and statistically justified correct standard errors (evaluation of the degree of uncertainty, so to speak) and correct confidence intervals for the degree of heavy-tailedness.

The chapter further provides applications of the inference approaches for the analysis of whether and how heavy-tailedness properties of emerging and developing markets such as markets for foreign exchange differ from those in developed economies. Finally, the chapter discusses recently developed general approaches to inference in economic and financial models under heterogeneity, dependence, and heavy-tailedness of largely unknown form.

Kazan, Russia   Marat Ibragimov
London, UK   Rustam Ibragimov
Berkeley, CA, USA   Johan Walden

Acknowledgements

We thank our co-authors Xavier Gabaix, Dwight Jaffee, Ulrich Müller, and Paul Kattuman for productive collaboration on joint projects presented in the manuscript, and very much hope that it will continue many years into the future. RI and JW began working on the topics of this book as Ph.D. students at the Department of Economics and School of Management at Yale University, respectively. They thank their advisors Donald Andrews, Peter Phillips, and Herbert Scarf (RI), and Zhiwu Chen, Will Goetzmann, and Jon Ingersoll (JW) for all their support and guidance. RI is also grateful to Donald Brown for inspiring research discussions while a student at Yale.

Rustam is indebted to MI for always being an example to follow and admire. He thanks his advisors Aydin Cecen and Shaturgun Sharakhmetov for support and research collaboration over many years. RI's work on majorization and probability inequalities was inspired by the Russian translation of Albert Marshall and Ingram Olkin' fundamental book on inequalities (Marshall and Olkin 1979) and the seminal paper by Proschan (1965) referred to therein. He is grateful to MI and Shaturgun Sharakhmetov for suggesting to read A. Marshall and I. Olkin' book many years ago while then a student at the Department of Mathematics at Tashkent State University. He thanks Victor de la Peña and Shaturgun Sharakhmetov for the long-time collaboration in research on inequalities in probability theory and to George Lentzas, Jingyuan Mo, and Artem Prokhorov for collaboration on the analysis of economic, financial, and econometric models under dependence and heavy-tailedness currently in progress.

We thank our colleagues at Yale University, Harvard University, University of California at Berkeley, Imperial College Business School, Innopolis University (Kazan, Russia), Kazan (Volga Region) Federal University, Tashkent State University, Stockholm School of Economics, and Uppsala University for inspiration and support. We also thank the participants at various seminars and conferences, where the results in the book were presented and discussed over several years, for many helpful comments. RI is indebted to Nail Bakirov (1952–2010) and Daniyar Mushtari (1945–2013) for inspiring discussions.

RI gratefully acknowledges that his work on many results presented in the book was supported by the Yale University Graduate Fellowship, the Cowles Foundation Prize, the US NSF grant SES-0820124, Clark and Warburg Research Funds (Department of Economics, Harvard University), and Harvard Academy Junior Faculty Development grant. RI and JW gratefully acknowledge that their joint research was supported by the National University of Singapore's Risk Management Institute. MI's work was supported by grants from the Economics Education and Research Consortium and the Global Development Network.

We are grateful to an anonymous reviewer for many useful comments and suggestions that greatly helped us to improve the manuscript. We also thank Eva Hiripi and her colleagues at Springer for their help in all stages of preparation of the manuscript for publication.

Contents

Chapter 1
Introduction

1.1 Background

The empirical and theoretical study of heavy-tailed distributions within economics and finance is by now a mature area of research, dating back more than 50 years. The first empirical study is usually attributed to Mandelbrot (1963), who noted that the changes of cotton prices seem to be well approximated by heavy-tailed so-called stable distributions. Loosely speaking, this means that rare events tend to happen much more often than they would if risk distributions had standard Gaussian (or other) thin tails. For example, the approximately 20 % drop of the stock market on the so-called Black Monday of October 19, 1987 would occur much less often than once in a billion years under standard assumptions of Gaussian distributions, and has been taken as evidence that stock market returns are heavy-tailed (see, for instance, the striking examples in Chap. 2 in Stock and Watson 2007 that illustrate inappropriateness of Gaussian distributions as models for financial returns based on their behavior during the Black Monday crisis).

In a theoretical study, Samuelson (1967b) formalized the intuition that when choosing a portfolio of risks, one should diversify, and thus not put all ones eggs in one basket, by showing that it is optimal for a risk averse expected utility optimizing agent to choose uniform diversification with equal weights for a portfolio of independent identically distributed (i.i.d.) risks, as long as these risks are not too heavy-tailed, in that they have finite second moments.[1] However, Samuelson (1967a) further noted that the optimality of diversification may not hold for extremely heavy-tailed distributions, an observation also made by Fama (1965b).

---

[1] In the terminology introduced by Rothschild and Stiglitz (1970), any other portfolio is riskier than the uniformly diversified one in that its distribution is a mean preserving spread of that of the uniformly diversified portfolio. We also note that independence of risks considered is crucial for the result, and cannot be replaced by the weaker condition of uncorrelated risks, as shown in Brumelle (1974).

© Springer International Publishing Switzerland 2015
M. Ibragimov et al., *Heavy-Tailed Distributions and Robustness in Economics and Finance*, Lecture Notes in Statistics 214, DOI 10.1007/978-3-319-16877-7_1

An example where the intuition breaks down is given, for instance, by the so-called Cauchy distributed risks. The Cauchy distribution belongs to the class of the above mentioned stable distributions studied by Mandelbrot (1963), and is defined through its probability density function

$$f(x) = \frac{1}{\pi(1 + x^2)},$$

or, equivalently, through its cumulative distribution function

$$F(x) = \frac{1}{2} + \frac{1}{\pi \arctan(x)}.$$

The risk of any portfolio of i.i.d. Cauchy distributed random variables (r.v.'s) will have the same distribution as that of an individual risk in the portfolio, so diversification is irrelevant for such risks (the reason for equality of the distributions is that a linear combination of i.i.d. Cauchy r.v.'s is again distributed as a Cauchy r.v., see Sect. 2.1.2).[2] Given the fundamental importance of risk diversification in many models of finance, economics, and insurance (including, e.g., in the celebrated Capital Asset Pricing Model), the conclusions that some real-world risk distributions are heavy-tailed, and that diversification may not be optimal for such distributions have potentially far-reaching consequences. Several challenges arise, however, when developing models for measuring the presence of heavy tails and understanding the consequences of their presence, which may explain why it has taken a long while for the field to gain momentum.

Heavy-tailedness is often defined in the context of power law distributions, so that for an r.v. (e.g., representing a risk, financial return or exchange rate) X,

$$P(X > x) \sim \frac{C_1}{x^{\zeta_1}}, \tag{1.1}$$

$$P(X < -x) \sim \frac{C_2}{x^{\zeta_2}}, \tag{1.2}$$

as $x \to +\infty$ (throughout the book, $f(x) \sim g(x)$ as $x \to +\infty$ means that $\lim_{x \to +\infty} \frac{f(x)}{g(x)} = 1$). Here, $\zeta_1, \zeta_2 > 0$, and $C_1, C_2 > 0$, are some constants. Relations (1.1) and (1.2) imply that

$$P(|X| > x) \sim \frac{C}{x^{\zeta}}, \tag{1.3}$$

---

[2]Throughout the book, the term "risk" is used as a synonym for the term "random variable," if this does not lead to a confusion. So that, here, for instance, we mean, in particular, that the risk (r.v. or loss) $\frac{1}{n}\sum_{i=1}^{n} Z_i$ of the portfolio of i.i.d. Cauchy risks (r.v.'s or losses) $Z_1, Z_2, \ldots, Z_n$ with equal weights has the same Cauchy distribution as does each of the r.v.'s Z_i.

with $\zeta = \min(\zeta_1, \zeta_2)$, and $C > 0$. The parameters ζ, ζ_1, and ζ_2 in (1.3) and (1.1)–(1.2) are referred to, respectively, as the tail index (or the tail exponent), the right tail index and the left tail index of the distribution of X. They characterize the heaviness (the rates of decay) of the tails of power law distributions (1.1)–(1.3). The more the probability mass in the tails, the smaller are the tail index parameters, and vice versa. Heavy-tailedness (i.e., the tail index ζ) of the variable X governs the likelihood of observing extreme fluctuations in the variable. The smaller values of the tail index ζ correspond to a higher degree of heavy-tailedness in X and, thus, to a larger likelihood of observing outliers and extreme fluctuations in realizations of this variable. The tail index may be regarded as being infinite: $\zeta = \infty$ for thin-tailed distributions like Gaussian or exponential ones.

A first challenge is that of empirically measuring whether a risk distribution is heavy-tailed and, if so, how heavy-tailed it is. In the rest of this chapter, we summarize the extensive literature that has documented the presence of heavy-tailedness with outliers, extreme observations and large fluctuations in many important variables in economics, finance, and insurance, and also further review several key properties of heavy-tailed and power law distributions. We then discuss several theoretical challenges, and introduce a simple framework that allows us to discuss the benefits and drawbacks of diversification in value at risk models under the presence of heavy tails. The framework will be further used, in subsequent chapters, to study robustness of a number of important models in economics and finance to heavy-tailedness assumptions.

Subsequently, the goal of Chap. 2 is to further emphasize importance of the analysis of heavy-tailedness in economic and financial markets. By focusing on a number of problems in different areas in economics, finance, risk management, and insurance, the results in the chapter demonstrate that the presence of heavy tails can either reinforce or reverse the implications of many important models in these fields, depending on the degree of heavy-tailedness. In particular, according to the results discussed in the chapter, the value $\zeta = 1$ is the dividing boundary between robustness and reversals of many economic and financial models in the case of heavy-tailed distributions.

1.2 Empirical Evidence on Heavy-Tailedness

Estimation of tail indices and inference on the degree of heavy-tailedness are inherently challenging, since extreme tail events—per definition—happen very rarely, and one therefore typically has a relatively small number of relevant observations, even in a large dataset. We will address this major challenge in Chap. 3. Below we summarize some studies that have analyzed the presence of heavy tails in economics, finance, risk management, and insurance.

Numerous contributions indicate that distributions of many key variables of interest in the aforementioned fields exhibit deviations from Gaussianity, including those in the form of heavy tails (see, among others, the discussion and reviews in

Cont 2001; Embrechts et al. 1997; Gabaix 2009; Ibragimov 2009a; Ibragimov and Walden 2007, and the references therein). This stream of literature goes back to Mandelbrot (1963) (see also Fama 1965a, and the papers in Mandelbrot 1997) who pioneered the study of heavy-tailed distributions in economics, finance, and other sciences. As was pointed out in many studies, the normal distribution paradigm does not hold in the case of financial returns, foreign exchange rates and many other variables of key interest in economics and finance that are increasingly prone to extreme behavior: e.g., as discussed in the previous section, according to the striking illustrations in Chap. 2 in Stock and Watson (2007), it is virtually impossible for Gaussian distributions to generate the extreme downfalls and large fluctuations in financial returns such as those observed during the 1987 Black Monday or other economic and financial crises.

Tail indices further characterize the maximal order of finite moments of the financial or economic variables X considered. The absolute moments of risks or returns X satisfying heavy-tailed power laws (1.1)–(1.3) are finite if and only if their order is less than $\zeta = \min(\zeta_1, \zeta_2) : E|X|^p < \infty$ if $p < \zeta$ and $E|X|^p = \infty$ if $p \geq \zeta$. In particular, the fourth moment of an r.v. X with a power law distribution is finite and, thus, its kurtosis is defined if and only if $\zeta > 4$. R.v.'s X that follow (1.1)–(1.3) have finite second moments $EX^2 < \infty$ (and, thus, well-defined variances) if and only if $\zeta > 2$. The first absolute moment of X in (1.1)–(1.3) is finite if and only if $\zeta > 1$. Examples of power laws (1.1) are given by stable and Student-t distributions. Further examples are provided by Singh–Maddala families and Pareto distributions (with equality in (1.1) for all $x > x_0$) that are widely used in income distribution modeling (see the discussion and references in Cowell and Flachaire 2007; Davidson and Flachaire 2007; Ibragimov 2009a). Important classes of heavy-tailed time series with power law distributions are provided by GARCH and stochastic volatility processes that have been used in many works in the literature for modeling a number of important economic and financial variables, including financial returns and foreign exchange rates (see Chap. 12 in Campbell et al. 1997; Cont 2001, and the references therein).

Many recent studies argue that the tail indices ζ in heavy-tailed models (1.3) typically lie in the interval $\zeta \in (2, 4)$ for financial returns and foreign exchange rates in developed economies (see, among others, Gabaix 2009; Gabaix et al. 2006; Ibragimov 2009a; Ibragimov and Walden 2007; Loretan and Phillips 1994a, and the references therein). These estimates imply that the above variables have finite variances and finite first moments; however, their fourth moments are infinite.

Heavy-tailed power law behavior is also exhibited by such important economic and financial variables as income and wealth (with $\zeta \in (1.5, 3)$ and $\zeta \approx 1.5$, respectively; see, among others, Gabaix 2009, and the references therein); city sizes and firm sizes (power laws with $\zeta \approx 1$ referred to as Zipf's law; see Axtell 2001; Gabaix 1999); financial returns from technological innovations, losses from operational risks and those from earthquakes and other natural disasters (with tail indices that can be considerably less than one, see Ibragimov et al. 2009; Nešlehova et al. 2006; Silverberg and Verspagen 2007, and Sect. 2.3.5.1 in this book, and references therein). Ozsoylev and Walden (2011) study asset pricing in markets where agents share information among themselves through a network, and specifically cover power law distributed networks with low tail indices.

The characteristics of heavy-tailedness such as tail indices in models (1.1)–(1.3) are of key interest for professionals in financial and insurance industries, risk managers, financial regulators, financial stability analysts, and policy makers concerned with the likelihood of large fluctuations or extreme values of financial returns, risks or foreign exchange rates, and the related risk measures. In particular, the estimates of the models are important in the analysis of loss exceedance probabilities and in assessing commonly used risk measures such as the value at risk and expected shortfall relatively far in the tails of heavy-tailed distributions considered.

Naturally, finiteness of variances for economic and financial indicators like financial returns and exchange rates is crucial for applicability of classical statistical and econometric approaches, including regression and least squares methods. In a similar fashion, the problem of potentially infinite fourth moments of economic and financial time series needs to be taken into account in applications of autocorrelation-based methods and related inference procedures in their analysis (see, among others, the discussion in Granger and Orr 1972, and in a number of more recent studies, e.g., Chap. 7 in Cont 2001; Davis and Mikosch 1998; Embrechts et al. 1997; Mikosch and Stărică 2000, and the references therein).

1.3 Diversification Under Heavy-Tailedness

How reasonable is it to use models with heavy-tailed distributions in economics and finance? We recount some objections against such modeling that may be made. One may argue that there are natural constraints on how large the changes in many of these variables can be. For example, the argument may be that the price of one unit of cotton could never increase beyond the point where the unit would cost more than the total amount of wealth in the world, the latter being bounded by the amount of resources on Earth. Consequently, decision rules that hold for all thin-tailed variables, the argument would go (as, e.g., laid out in Samuelson 1967b), would then hold in general.

Another objection that is related to the previous one, and applicable to many financial markets, is that even if real risk distributions are heavy-tailed, usually contracts come with some type of limited liability. For example, the owners of a publicly traded firm that faces heavy-tailed operational risks are not necessarily exposed to such heavy tails, because firms in the USA have limited liability, bounding the potential downside. Again, this leads to a bound on losses. Other examples include explicit and implicit government guarantees, e.g., in banking and insurance.

One may argue that seemingly heavy-tailed risk distributions arise because of our limited knowledge about the true underlying structure of the risks under study, and that a model that takes this structure into account may be able to correct for extremes without assuming heavy tails. That is, with additional information about the structure of uncertainty, risk distributions may not be heavy-tailed. As an example, consider a stochastic process that is normally distributed at each point in time, but with stationary random volatility in each period, so that its unconditional distribution is heavy-tailed. Conditioning on full information about the current volatility, the distribution is thin-tailed and normally distributed. But without any information about current volatility, the unconditional distribution must be used, which is heavy-tailed. Heavy-tailedness thus arises because of lack of information. This argument may indeed suggest that future—better—models will mitigate the need for using heavy-tailed risks distributions. However, until such superior models are developed, we will have to rely on models with heavy-tailed risks. We think it is unlikely that drastic improvements in the modeling of, for example, catastrophic events will mitigate the need for using heavy-tailed risk distributions in the near future.

Finally, there is a theoretical question of in what sense one can say that diversification increases riskiness when tails are heavy. For thin-tailed and moderately heavy-tailed risks, the theory in Rothschild and Stiglitz (1970) provides a tight link between portfolio riskiness and the utility of agents: The distributions of individual risks are less concentrated around their mean than the risk distributions of diversified portfolios of i.i.d. risks with equal weights, and therefore inferior to any risk-averse expected utility optimizing agent. For extremely heavy-tailed risks, however, the expected utility argument may break down since expected values of utility functions may not be defined in this case. Consequently, a justification for the meaning of increased risk is needed.

In the next chapter, we will introduce a theory that is applied to a large class of risks and their portfolios and is further used, in subsequent chapters, to study robustness of a number of important economic and financial models to heavy-tailedness. To keep things simple and intuitive, however, we first consider specific risk distributions in this section, and introduce a simple framework that allows us to discuss these theoretical concerns.

We focus on the (reflected) Lévy distribution, which is extremely heavy-tailed, with the tail index of 0.5. The pdf of the distribution is

$$f(x) = \begin{cases} \sqrt{\frac{\sigma}{2\pi}} e^{-\sigma/2(\mu-x)} (\mu-x)^{-3/2}, & x \le \mu, \\ 0, & x > \mu, \end{cases}$$

where $\mu > 0$ and $\sigma > 0$ are, respectively, the location and spread parameters. Its cdf is given by

$$F(x) = \begin{cases} \text{Erf}\left(\frac{\sigma}{\sqrt{2(\mu-x)}}\right), & x \le \mu, \\ 1, & x > \mu. \end{cases} \tag{1.4}$$

Here, $\text{Erf}(y) = \frac{2}{\sqrt{\pi}} \int_0^y e^{-t^2} dt$ is the error function; see Abramowitz and Stegun (1970). We denote by $L_{\mu,\sigma}$ the class of r.v.'s with the above Lévy distributions and

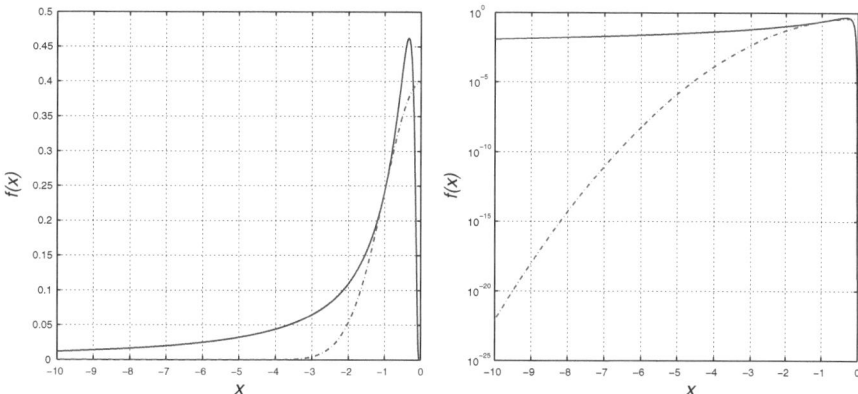

Fig. 1.1 P.d.f. of heavy-tailed Lévy distribution (*solid*) and thin-tailed normal distribution (*dotted*)

write $X \sim L_{\mu,\sigma}$ if the distribution of the r.v. X belongs to the class. It is easy to see that r.v.'s $X \sim L_{\mu,\sigma}$ are extremely heavy-tailed: The tails of their distributions satisfy power law relations (1.2)–(1.3) with the tail index $\zeta = 0.5 < 1$. In Fig. 1.1, we show the p.d.f. of a risk with the above heavy-tailed Lévy distribution and that of a risk with the standard normal distribution. Since the Lévy distribution is one-sided, we focus on the negative half-line. As seen in the left panel, the p.d.f. of the normal distribution is indistinguishable from zero for $x < -5$, whereas it is significantly different from zero for the Lévy distribution, and only slowly decreasing beyond $x < -5$. In the right panel, the decay is shown in logarithmic scale. Again, the difference between the two distributions is striking: The normal distribution decreases faster than linearly as x decreases, whereas the Lévy distribution seems to be almost flat.

Given that negative values of x represent losses, it is clear that a larger value of the spread parameter, σ, for a Lévy distribution implies a worse situation (for a fixed μ), using any reasonable definition. In fact, as is seen from Eq. (1.4), increasing σ for a loss variable $X \sim L_{\mu,\sigma}$ leads to a loss whose magnitude (the absolute value) dominates that of X in the sense of first-order stochastic dominance. Here, a risk X_2 (e.g., the loss magnitude) is said to first-order stochastically dominate X_1 if $P(X_2 > x) \geq P(X_1 > x)$ for all x: So that, in particular, the magnitude of the loss X_2 is more likely to be larger than for X_1. If the inequality is strict for some x, the dominance is said to be strict. Equivalently, first-order stochastic dominance can be defined in terms of cdf's F_1 and F_2 of the risks X_1 and X_2 : $F_2(x) \leq F_1(x)$ for all x. Clearly, first-order stochastic dominance is a very strong concept. The only behavioral assumption needed for an expected utility maximizing agent to prefer risk X_1 over X_2 is that she prefers smaller losses to larger losses.

Diversification of losses in the Lévy class $L_{\mu,\sigma}$ increases the distribution spread σ. This follows from the fact that distributions of such losses are particular cases of stable distributions considered in the next chapter. Stable distributions are closed under portfolio formation and, in particular, for i.i.d. losses $X_1, X_2 \sim L_{\mu,\sigma}$, the loss $Z = (X_1 + X_2)/2$ from their portfolio with equal weights also belongs to the same class: $Z \sim L_{\mu,2\sigma}$ but with the spread parameter of 2σ. This follows from the following uniform diversification rule for portfolios of n i.i.d. losses from the class $L_{\mu,\sigma}$ with equal weights (here and throughout, for risks or losses $Z_1, Z_2, \ldots, Z_n$, as usual, we denote by $\overline{Z}_n = (1/n) \sum_{i=1}^{n} Z_i$ their sample mean, that is, the risk or loss of the most diversified portfolio of Z_i's with equal weights):

$$X_i \in L_{\mu,\sigma}, \ i = 1, \ldots, n \quad \Longrightarrow \quad \overline{X}_n \in L_{\mu,n\sigma},$$

or, equivalently,

$$X_i \in L_{\mu,\sigma}, \ i = 1, \ldots, n \quad \Longrightarrow \quad \overline{X}_n =^d nX_1, \tag{1.5}$$

where, throughout the book, for two r.v.'s Y and Z, we write $Y =^d Z$ if Y and Z have the same distribution (see also relation (2.3) for general stable distributions in the next chapter). Therefore, uniform diversification of n i.i.d losses from the extremely heavy-tailed class $L_{\mu,\sigma}$ increases the spread parameter from σ to $n\sigma$ in line with the example above, and is therefore sub-optimal.

Now consider a financial market in which there are n publicly traded firms. Each firm has limited liability, so that if it generates profits X, the value of the firm to shareholders is $I(X)$, where I is the indicator function

$$I(x) = \begin{cases} x, \ x \geq 0, \\ 0, \ x < 0. \end{cases}$$

Consider first the situation, where each firm, i, has invested in one $L_{\mu,\sigma}$ distributed risk, X_i, where we for simplicity assume that μ and σ are the same across firms, and the risks are i.i.d. The distribution of $I(X_i)$ is, of course, thin-tailed, being bounded below by 0, so the classical results on diversification of thin-tailed risks apply, and a portfolio investor will choose a uniformly diversified portfolio when investing in the market.

Now consider instead the situation where uniform diversification occurs within the firm, in which case one firm chooses to allocate its investment equally across the multiple risks. In this case, the firms choose a portfolio with equal weights and the risk $\overline{X}_n$, and the payout to a shareholder is then $I(\overline{X}_n)$. Now it is clear from the previous argument on diversification of heavy-tailed risks that X_1 strictly first order dominates $\overline{X}_n$, and since the indicator function preserves first-order stochastic dominance, it is optimal for the firm to only invest in one risk, i.e., any diversification is suboptimal.

From these two examples we see that limited liability does not necessarily restore the optimality of diversification: it depends on whether the limited liability operator acts on the individual risks, as in the first example, or on the portfolio of risks, as in the second. Note that the second example is extremely robust to assumptions about agents' preferences: They could be risk-neutral, risk-averse, etc.; any expected utility maximizing agent who prefers more to less will agree that the least diversified outcome is optimal.

The second example also shows that these results are robust even if the risks are not heavy-tailed asymptotically. Indeed, it is straightforward to show that similar results are obtained for risks Y_i, which have the same pdf as Lévy distributed risks until some negative point far out in the domain, $x \ll 0$, but then the tails of their distributions decrease faster. For such risks, $I\left(\overline{Y}_n\right)$ will have a distribution very similar to that of $I\left(\overline{X}_n\right)$. The optimal outcome is therefore very close to the one for X-risks: Again, the optimal decision is basically to avoid any type of diversification.

This example demonstrates that it is indeed possible to compare portfolio riskiness for heavy-tailed risks, that sub-optimality of diversification may also hold in situations of limited liability, and that the results may also hold in situations where the risks are not heavy-tailed asymptotically. In the next chapter, we expand our analysis to a much broader class of risks, and then study in detail a number of important models in economics, finance, and insurance, where heavy-tailedness may lead to radically different outcomes than what classical theory predicts.

Chapter 2
Implications of Heavy-Tailedness

2.1 Diversification Analysis Via Majorization

2.1.1 Majorization Relation

This chapter demonstrates how majorization theory provides a powerful tool for the study of robustness of many important models in economics, finance, econometrics, statistics, risk management, and insurance to heavy-tailedness assumptions. The majorization relation is a formalization of the concept of diversity in the components of vectors. Over the past decades, majorization theory, which focuses on the study of this relation and functions that preserve it, has found applications in disciplines ranging from statistics, probability theory, and economics to mathematical genetics, linear algebra, and geometry (see Marshall et al. 2011, and the references therein).

A vector $v \in \mathbf{R}^n$ is said to be majorized by a vector $w \in \mathbf{R}^n$, written $v \prec w$, if $\sum_{i=1}^{k} v_{[i]} \leq \sum_{i=1}^{k} w_{[i]}$, $k = 1, \ldots, n-1$, and $\sum_{i=1}^{n} v_{[i]} = \sum_{i=1}^{n} w_{[i]}$, where $v_{[1]} \geq \ldots \geq v_{[n]}$ and $w_{[1]} \geq \ldots \geq w_{[n]}$ denote components of v and w in decreasing order. The relation $v \prec w$ implies that the components of the vector v are less diverse than those of w. In this context, it is easy to see that the following relations hold:

$$\left(1/n, \ldots, 1/n \right) \prec (w_1, \ldots, w_n) \prec \left(1, 0, \ldots, 0 \right),$$

for all $w_i \geq 0$ such that $\sum_{i=1}^{n} w_i = 1$, and

$$(1/(n+1), \ldots, 1/(n+1), 1/(n+1)) \prec (1/n, \ldots, 1/n, 0), \quad n \geq 1.$$

A function $\phi : A \to \mathbf{R}$ defined on $A \subseteq \mathbf{R}^n$ is called *Schur-convex* (resp., *Schur-concave*) on A if $(v \prec w) \Longrightarrow (\phi(v) \leq \phi(w))$ (resp. $(v \prec w) \Longrightarrow (\phi(v) \geq \phi(w))$ for all $v, w \in A$. If, in addition, $\phi(v) < \phi(w)$ (resp., $\phi(v) > \phi(w)$) whenever $v \prec w$

© Springer International Publishing Switzerland 2015

M. Ibragimov et al., *Heavy-Tailed Distributions and Robustness in Economics and Finance*, Lecture Notes in Statistics 214, DOI 10.1007/978-3-319-16877-7_2

and v is not a permutation of w, then ϕ is said to be *strictly* Schur-convex (resp., *strictly* Schur-concave) on A.

Examples of strictly Schur-convex functions $\phi : \mathbf{R}_+^n \to \mathbf{R}$ are given by $\phi(w_1, \ldots, w_n) = \sum_{i=1}^n w_i^2$ and, more generally, by $\phi_p(w_1, \ldots, w_n) = \sum_{i=1}^n w_i^p$ for $p > 1$. The functions $\phi_p(w_1, \ldots, w_n)$ are strictly Schur-concave for $p < 1$ (see Proposition 3.C.1.a in Marshall et al. 2011).[1]

Denote $\mathcal{I}_n = \{w = (w_1, \ldots, w_n) \in \mathbf{R}_+ : \sum_{i=1}^n w_i = 1\}$. Consider two vectors of portfolio weights $v, w \in \mathcal{I}_n$. Further, denote $\underline{w} = (1/n, 1/n, \ldots, 1/n) \in \mathcal{I}_n$ and $\overline{w} = (1, 0, \ldots, 0) \in \mathcal{I}_n$.

If $v \prec w$, it is natural to think about the portfolio with weights v as being more diversified than that with weights w. That is, for example, the portfolio with equal weights $\underline{w}$ is the most diversified and the portfolio with weights $\overline{w}$ consisting of one risk is the least diversified among all the portfolios with weights $w \in \mathcal{I}_n$ (in this regard, the notion of one portfolio being more or less diversified than another one is, in some sense, the opposite of the majorization ordering for vectors of weights for the portfolio).

In what follows, given a loss probability $q \in (0, 1)$ and a r.v. (risk) X we denote by $\text{VaR}_q(X)$ the value at risk (VaR) of X at level q, that is, the negative of its q-quantile: $\text{VaR}_q(X) = -\inf\{z \in \mathbf{R} : P(X \le z) > q\}$ (see, among others, Artzner et al. 1999, Chap. 2 in McNeil et al. 2005, and Chap. 1 in Christoffersen 2012). For an r.v. X with a strictly increasing cdf $F(z) = P(X \le z)$ one thus has $\text{VaR}_q(X) = -F^{-1}(q)$. That is, as discussed, for example, in Sect. 1.9 in Christoffersen (2012), the value at risk $\text{VaR}_q(X)$ is defined as the number such that the risk X results in a worse loss only with probability q.[2]

Let $\mathcal{X}$ be a certain linear space of r.v.'s X defined on a probability space $(\Omega, \Im, P)$. We assume that $\mathcal{X}$ contains all degenerate r.v.'s $X \equiv a \in \mathbf{R}$. According to the definition in Artzner et al. (1999) (see also Frittelli and Gianin 2002; McNeil et al. 2005), a functional $\mathcal{R} : \mathcal{X} \to R$ is said to be a *coherent* measure of risk if it satisfies the following axioms:

a1. (Monotonicity) $\mathcal{R}(X) \ge \mathcal{R}(Y)$ for all $X, Y \in \mathcal{X}$ such that $X \le Y$ (a.s.), that is, $P(X \le Y) = 1$.
a2. (Translation invariance) $\mathcal{R}(X + a) = \mathcal{R}(X) - a$ for all $X \in \mathcal{X}$ and any $a \in \mathbf{R}$.
a3. (Positive homogeneity) $\mathcal{R}(\lambda X) = \lambda \mathcal{R}(X)$ for all $X \in \mathcal{X}$ and any $\lambda \ge 0$.
a4. (Subadditivity) $\mathcal{R}(X + Y) \le \mathcal{R}(X) + \mathcal{R}(Y)$ for all $X, Y \in \mathcal{X}$.

---

[1] The functions ϕ_p have the same form as measures of diversification considered in Bouchaud and Potters (2004), Chap. 12, p. 205.

[2] Throughout the chapter, we interpret the negative values of risks X as a risk holder's losses. This interpretation is similar to that in Artzner et al. (1999) and Christoffersen (2012) and is in contrast to Chap. 2 in McNeil et al. (2005) who interpret positive values of risks X as losses. All the results presented and discussed in the chapter can be easily reformulated in terms of interpretation of positive values of the risks as losses (see Ibragimov 2009a,b, for details).

In some papers (see Fölmer and Schied 2002; Frittelli and Gianin 2002), the axioms of positive homogeneity and subadditivity are replaced by the following weaker axiom of convexity:

a5. (Convexity) $\mathcal{R}(\lambda X + (1 - \lambda)Y) \leq \lambda \mathcal{R}(X) + (1 - \lambda)\mathcal{R}(Y)$ for all $X, Y \in \mathcal{X}$ and any $\lambda \in [0, 1]$

(clearly, a5 follows from a3 and a4).

It is easy to verify that the value at risk $\mathrm{VaR}_q(X)$ satisfies the axioms of monotonicity, translation invariance, and positive homogeneity a1, a2, and a3. However, as follows from the counterexamples constructed by Artzner et al. (1999) and McNeil et al. (2005), in general, it fails to satisfy the subadditivity and convexity properties a4 and a5 (see Remarks 2.1.1 and 2.1.2 for implications of the results in this section for coherency of the VaR and the asymptotic analysis in the case of distributions with regularly varying heavy tails).

Consider risks $X_1, \ldots, X_n$. For $w = (w_1, \ldots, w_n) \in \mathcal{I}_n$, we denote by Z_w the risk of the portfolio of X_i's with weights w. The expressions $\mathrm{VaR}_q(Z_{\underline{w}})$ and $\mathrm{VaR}_q(Z_{\overline{w}})$ are, thus, the values at risk of the (most diversified) portfolio with equal weights $\underline{w}$ and of the (least diversified) portfolio with weights $\overline{w}$ that consists of only one return (risk).

In order to formulate the main results of the chapter on the effects of heavy-tailedness on diversification analysis in the VaR framework, we need to introduce several classes of distributions that will be dealt with throughout the book.

2.1.2   Notation and Classes of Distributions

In what follows, a univariate density $f(x)$, $x \in \mathbf{R}$, will be referred to as symmetric (about zero) if $f(x) = f(-x)$ for all $x > 0$. In addition, as usual, an absolutely continuous distribution or an r.v. X with the density $f(x)$ will be called symmetric if $f(x)$ is symmetric (about zero).

An r.v. X with density $f(x)$, $x \in \mathbf{R}$, and the convex support $\Omega = \{x \in \mathbf{R} : f(x) > 0\}$ is log-concavely distributed if $log\ f(x)$ is concave in $x \in \Omega$, that is, if for all $x_1, x_2 \in \Omega$, and any $\lambda \in [0, 1]$, $f(\lambda x_1 + (1 - \lambda)x_2) \geq (f(x_1))^\lambda (f(x_2))^{1-\lambda}$ (see An 1998; Bagnoli and Bergstrom 2005, and Sect. 18.B in Marshall et al. 2011). Examples of log-concave distributions include the normal distribution, the uniform density, the exponential density, the Gamma density $f(x) = \lambda^r x^{r-1} e^{-\lambda x} / \Gamma(r)$, $x \geq 0$, with the shape parameter $r \geq 1$; the Beta distribution with the density $f(x) = [B(a, b)]^{-1} x^{a-1}(1 - x)^{b-1}$, $0 \leq x \leq 1$ for $a \geq 1$ and $b \geq 1$; and the Weibull density $f(x) = \lambda \xi exp(\lambda x + \xi - \xi e^{\lambda x})$ with the shape parameter $\xi \geq 1$.[3] The class of log-concave distributions is closed under convolution. Log-concave distributions have many other appealing properties that have been utilized in a number of works in

---

[3]Here, as usual, $\Gamma(r)$ and $B(a, b)$ denote the Gamma and Beta functions.

economics and finance (see the surveys in An 1998; Bagnoli and Bergstrom 2005; Karlin 1968; Marshall et al. 2011). However, such distributions cannot be used in the study of heavy-tailedness phenomena since any log-concave density is extremely thin-tailed: In particular, if a r.v. X is log-concavely distributed, then its density has at most an exponential tail, that is, $f(x) = O(exp(-\lambda x))$ for some $\lambda > 0$, as $x \to \infty$ and all the power moments $E|X|^p, p > 0$, of the r.v. are finite (see Corollary 1 in An 1998). Throughout the monograph, $\mathcal{LC}$ denotes the class of symmetric log-concave distributions ($\mathcal{LC}$ stands for "log-concave").

For $0 < \alpha \leq 2$ and $\sigma > 0$, we denote by $S_\alpha(\sigma)$ the symmetric stable distribution with the index of stability (the characteristic exponent) α and the scale parameter σ. That is, $S_\alpha(\sigma)$ is the distribution of an r.v. X with the characteristic function (cf) $E(e^{ixX}) = exp\{-\sigma^\alpha |x|^\alpha\}, i^2 = -1, x \in \mathbf{R}$. The distribution $S_\alpha(\sigma)$ is a particular (symmetric) case of general stable distributions $S_\alpha(\sigma, \beta, \mu)$ with the parameterizations for cf's that involve, in addition to α and σ, the skewness parameter β and the location parameter μ (see Ibragimov 2009b; Uchaikin and Zolotarev 1999; Zolotarev 1986) The distribution $S_\alpha(\sigma)$ is symmetric about the location parameter $\mu = 0$ and has the skewness parameter $\beta = 0$. In the case $\beta \neq 0$, the stable distributions are asymmetric. For instance, the stable reflected Lévy distribution considered in Sect. 1.3 is one-sided and is concentrated on the interval $(-\infty, \mu]$; it has the skewness parameter $\beta = -1$. Similarly, stable distributions with the skewness parameter $\beta = 1$ are one-sided and concentrated on the interval $[\mu, \infty)$. In what follows, we write $X \sim S_\alpha(\sigma)$, if the r.v. X has the stable distribution $S_\alpha(\sigma)$.

A closed form expression for the density $f(x)$ of the general stable distribution $S_\alpha(\sigma, \beta, \mu)$ is available in the following cases (and only in those cases): $\alpha = 2$ that corresponds to Gaussian distributions; $\alpha = 1$ and $\beta = 0$ for Cauchy distributions with densities

$$f(x) = \frac{\sigma}{\pi(\sigma^2 + (x - \mu)^2)}; \tag{2.1}$$

$\alpha = 1/2$ and $\beta = \pm 1$ for Lévy distributions with densities

$$f(x) = \left(\frac{\sigma}{2\pi}\right)^{1/2} exp\left(-\frac{\sigma}{2x}\right) x^{-3/2}, \tag{2.2}$$

$x \geq 0; f(x) = 0, x < 0$, where $\sigma > 0$, and their shifted and reflected versions as in the example considered in Sect. 1.3. Degenerate distributions correspond to the limiting case with $\sigma = 0$.

The index of stability α characterizes the heaviness (the rate of decay) of the tails of stable distributions $S_\alpha(\sigma)$. In particular, if $X \sim S_\alpha(\sigma), 0 < \alpha < 2$, then the distribution of X satisfies power law (1.1)–(1.3) with the tail index ζ equal to the index of stability $\alpha : \zeta = \alpha$. In other words, all stable distributions except Gaussian ones with the index of stability $\alpha = 2$ are heavy-tailed with the tail index $\zeta = \alpha$. This implies that the p-th absolute moments $E|X|^p$ of a r.v. $X \sim S_\alpha(\sigma), \alpha \in (0, 2)$

are finite if $p < \alpha$ and are infinite otherwise. In particular, the second moments of non-Gaussian stable distributions are infinite and, thus, their variances and standard deviations are not defined. In the case $\alpha > 1$ the location parameter $\mu = 0$ is the mean of the distribution $S_\alpha(\sigma)$. The scale parameter σ is a generalization of the concept of standard deviation; it coincides with the standard deviation divided by $\sqrt{2}$ in the special case of Gaussian distributions ($\alpha = 2$).

Stable distributions are closed under portfolio formation. In particular, if $X_i \sim S_\alpha(\sigma)$ are i.i.d. symmetric stable risks, then, for all portfolio weights $w_i \geq 0$, $i = 1, \ldots, n$,

$$\sum_{i=1}^{n} w_i X_i =^d \left(\sum_{i=1}^{n} w_i^\alpha \right)^{1/\alpha} X_1, \tag{2.3}$$

or, equivalently, $\sum_{i=1}^{n} w_i X_i \sim S_\alpha(\tilde{\sigma})$, where $\tilde{\sigma} = \sigma \left(\sum_{i=1}^{n} w_i^\alpha \right)^{1/\alpha}$ (see McNeil et al. 2005; Rachev and Mittnik 2000; Uchaikin and Zolotarev 1999; Zolotarev 1986, for a review of properties of stable distributions).

Denote by $\overline{\mathcal{CS}}$ the class of distributions which are convolutions of symmetric stable distributions $S_\alpha(\sigma)$ with the indices of stability $\alpha \in (1, 2]$ and $\sigma > 0$ (here and below, $\mathcal{CS}$ stands for "convolutions of stable"; the overline indicates that convolutions of stable distributions with indices of stability *greater* than the threshold value $\alpha = 1$ are taken). That is, $\overline{\mathcal{CS}}$ consists of distributions of r.v.'s X such that, for some $k \geq 1$, $X = Y_1 + \ldots + Y_k$, where $Y_i \sim S_{\alpha_i}(\sigma_i)$, $i = 1, \ldots, k$, are independent stable r.v.'s with $\alpha_i \in (1, 2]$, $\sigma_i > 0$, $i = 1, \ldots, k$.

Further, $\underline{\mathcal{CS}}$ stands for the class of distributions which are convolutions of symmetric stable distributions $S_\alpha(\sigma)$ with indices of stability $\alpha \in (0, 1)$ and $\sigma > 0$ (the underline indicates considering stable distributions with indices of stability *less* than the threshold value $\alpha = 1$). That is, $\underline{\mathcal{CS}}$ consists of distributions of r.v.'s X such that, for some $k \geq 1$, $X = Y_1 + \ldots + Y_k$, where $Y_i \sim S_{\alpha_i}(\sigma_i)$, $i = 1, \ldots, k$, are independent stable r.v.'s with $\alpha_i \in (0, 1)$, $\sigma_i > 0$, $i = 1, \ldots, k$.

Finally, we denote by $\overline{\mathcal{CSLC}}$ the class of convolutions of distributions from the classes $\overline{\mathcal{CS}}$ and $\mathcal{LC}$. That is, $\overline{\mathcal{CSLC}}$ is the class of convolutions of symmetric distributions which are either log-concave or stable with indices of stability greater than one ($\mathcal{CSLC}$ is the abbreviation of "convolutions of stable and log-concave"). In other words, $\overline{\mathcal{CSLC}}$ consists of distributions of r.v.'s X such that $X = Y_1 + Y_2$, where Y_1 and Y_2 are independent r.v.'s with distributions belonging to $\overline{\mathcal{CS}}$ or $\mathcal{LC}$.

All the classes $\mathcal{LC}$, $\overline{\mathcal{CSLC}}$, $\overline{\mathcal{CS}}$, and $\underline{\mathcal{CS}}$ are closed under convolutions. In particular, the class $\overline{\mathcal{CSLC}}$ coincides with the class of distributions of r.v.'s X such that, for some $k \geq 1$,

$$X = Y_1 + \ldots + Y_k, \tag{2.4}$$

where Y_i, $i = 1, \ldots, k$, are independent r.v.'s with distributions belonging to $\overline{\mathcal{CS}}$ or $\mathcal{LC}$.

The distributions of r.v.'s X in $\overline{\mathcal{CSLC}}$ are moderately heavy-tailed in the sense that, as is easy to see, as $x \to \infty$, their tails $P(|X| > x)$ decay to zero faster than those of power law distributions in (1.3) with the tail index $\zeta = 1$. In addition, an r.v. $X \sim \overline{\mathcal{CSLC}}$ follows power laws (1.1)–(1.3) with the tail indices $\zeta_1 = \zeta_2 = \zeta \in (1, 2)$ if its convolution representation (2.4) includes at least one r.v. Y_i from the class $\overline{\mathcal{CS}}$. This implies the distributions of r.v.'s $X \sim \overline{\mathcal{CSLC}}$ have finite absolute first moments: $E|X| < \infty$.

In contrast, the distributions of r.v.'s X from the class $\underline{\mathcal{CS}}$ are extremely heavy-tailed in the sense that they follow power laws (1.1)–(1.3) with the tail indices $\zeta_1 = \zeta_2 = \zeta \in (0, 1)$. Thus, the absolute first moments of the r.v.'s are infinite: $E|X| = \infty$.

Cauchy distributions $S_1(\sigma)$ that follow power laws (1.1)–(1.3) with the tail index $\zeta = 1$ are at the dividing boundary between the classes $\underline{\mathcal{CS}}$ and $\overline{\mathcal{CS}}$ (and between the classes $\underline{\mathcal{CS}}$ and $\overline{\mathcal{CSLC}}$).

As follows from stability property (2.3), a linear combination of independent stable r.v.'s with the *same index of stability* α also has a stable distribution with the same α. However, in general, this does not hold in the case of convolutions of stable distributions with *different* indices of stability. Therefore, the class $\overline{\mathcal{CS}}$ of *convolutions* of symmetric stable distributions with *different* indices of stability $\alpha \in (1, 2]$ is wider than the class of *all* symmetric stable distributions $S_\alpha(\sigma)$ with $\alpha \in (1, 2]$ and $\sigma > 0$. Similarly, the class $\underline{\mathcal{CS}}$ is wider than the class of *all* symmetric stable distributions $S_\alpha(\sigma)$ with $\alpha \in (0, 1)$ and $\sigma > 0$.

Clearly, $\overline{\mathcal{CS}} \subset \overline{\mathcal{CSLC}}$ and $\mathcal{LC} \subset \overline{\mathcal{CSLC}}$. It should also be noted that the class $\overline{\mathcal{CSLC}}$ is wider than the class of (two-fold) convolutions of log-concave distributions with stable distributions $S_\alpha(\sigma)$ with $\alpha \in (1, 2]$ and $\sigma > 0$.

In what follows, we write $X \sim \mathcal{LC}$ (resp., $X \sim \overline{\mathcal{CSLC}}, X \sim \overline{\mathcal{CS}}$ or $X \sim \underline{\mathcal{CS}}$) if the distribution of the r.v. X belongs to the class $\mathcal{LC}$ (resp., $\overline{\mathcal{CSLC}}, \overline{\mathcal{CS}}$ or $\underline{\mathcal{CS}}$).

2.1.3 Diversification in Value at Risk Models for Heavy-Tailed Risks[4]

A simple example where diversification is preferable is provided by the standard case with normal risks. Let $n \geq 2$, $q \in (0, 1/2)$, and let $X_1, \ldots, X_n \sim S_2(\sigma, 0, 0)$ be i.i.d. symmetric normal r.v.'s. Then, for the portfolio of X_i's with the equal weights

---

[4]This section draws upon material from the following articles:

Ibragimov (2009b) "Portfolio diversification and VaR under thick-tailedness", *Quantitative Finance*, Vol. 9, No. 5, 565–580, and

Ibragimov (2009a) "Heavy-tailed densities," "The New Palgrave Dictionary of Economics," Eds. Steven N. Durlauf and Lawrence E. Blume, Palgrave Macmillan, reproduced with permission of Palgrave Macmillan. The full published version of this publication is available from: http://www.dictionaryofeconomics.com/article?id=pde2009_H000191.

$\underline{w} = (1/n, 1/n, \ldots, 1/n)$ we have $Z_{\underline{w}} = (1/n) \sum_{i=1}^{n} X_i =^d (1/\sqrt{n}) X_1$. Consequently, by positive homogeneity of the VaR, $\mathrm{VaR}_q(Z_{\underline{w}}) = (1/\sqrt{n}) \mathrm{VaR}_q(X_1) = (1/\sqrt{n}) \mathrm{VaR}_q(Z_{\overline{w}}) < \mathrm{VaR}_q(Z_{\overline{w}})$. That is, the VaR of the most diversified portfolio with equal weights $\underline{w}$ is less than that of the least diversified portfolio with weights $\overline{w}$ consisting of only one risk Z_1.

Theorem 2.1.1 shows that similar results also hold for all moderately heavy-tailed risks X_i with arbitrary weights $w = (w_1, \ldots, w_n) \in \mathbf{R}_+^n$.[5] In all these settings, diversification of a portfolio of $X_i's$ leads to a decrease in the VaR of its return $Z_w = \sum_{i=1}^{n} w_i X_i$.

Theorem 2.1.1 *Let $q \in (0, 1/2)$ and let X_i, $i = 1, \ldots, n$, be i.i.d. risks such that $X_i \sim \overline{\mathcal{CSLC}}$, $i = 1, \ldots, n$. Then*

(i) *$\mathrm{VaR}_q(Z_v) < \mathrm{VaR}_q(Z_w)$ if $v \prec w$ and v is not a permutation of w (in other words, the function $\psi(w, q) = \mathrm{VaR}_q(Z_w)$ is strictly Schur-convex in $w \in \mathbf{R}_+^n$).*

(ii) *In particular, $\mathrm{VaR}_q(Z_{\underline{w}}) < \mathrm{VaR}_q(Z_w) < \mathrm{VaR}_q(Z_{\overline{w}})$ for all $q \in (0, 1/2)$ and all weights $w \in \mathcal{I}_n$ such that $w \neq \underline{w}$ and w is not a permutation of $\overline{w}$.*

Let us illustrate the settings where diversification is suboptimal in the VaR framework. Let $q \in (0, 1)$ and let $X_1, \ldots, X_n$ be i.i.d. risks with a stable reflected Lévy distribution with $\mu = 0$ considered in Sect. 1.3. Using (1.5) for the portfolio of $X_i's$ with equal weights $w_i = 1/n$, we get $Z_{\underline{w}} = (1/n) \sum_{i=1}^{n} X_i =^d n X_1$. Consequently, by positive homogeneity of the VaR, $\mathrm{VaR}_q(Z_{\underline{w}}) = n \mathrm{VaR}_q(X_1) = n \mathrm{VaR}_q(Z_{\overline{w}}) > \mathrm{VaR}_q(Z_{\overline{w}})$. Thus, the VaR of the least diversified portfolios with weights $\overline{w}$ that consists of only one risk is less than the VaR of the most diversified portfolio with equal weights $\underline{w}$.

Theorem 2.1.2 shows that similar conclusions hold for portfolio VaR comparisons with arbitrary weights $w = (w_1, \ldots, w_n) \in \mathbf{R}_+^n$ under the general assumption that the distributions of the risks $X_1, \ldots, X_n$ are extremely heavy-tailed. In such settings, the results in Theorem 2.1.1 are reversed and diversification of a portfolio of the risks X_i increases the VaR of its return.

Theorem 2.1.2 *Let $q \in (0, 1/2)$ and let X_i, $i = 1, \ldots, n$, be i.i.d. risks such that $X_i \sim \mathcal{CS}$, $i = 1, \ldots, n$. Then*

(i) *$\mathrm{VaR}_q(Z_v) > \mathrm{VaR}_q(Z_w)$ if $v \prec w$ and v is not a permutation of w (in other words, the function $\psi(w, q) = \mathrm{VaR}_q(Z_w)$, is strictly Schur-concave in $w \in \mathbf{R}_+^n$).*

(ii) *In particular, $\mathrm{VaR}_q(Z_{\overline{w}}) < \mathrm{VaR}_q(Z_w) < \mathrm{VaR}_q(Z_{\underline{w}})$ for all $q \in (0, 1/2)$ and all weights $w \in \mathcal{I}_n$ such that $w \neq \underline{w}$ and w is not a permutation of $\overline{w}$.*

---

[5]In particular, the results Theorems 2.1.1 and 2.1.2 and their analogues under dependence provided by Theorems 5.1 and 5.2 in Ibragimov (2009b) substantially generalize the riskiness analysis for uniform (equal weights) portfolios of independent stable risks considered, among others, in the papers by Fama (1965b), Samuelson (1967a) and Ross (1976): These theorems demonstrate that the formalization of portfolio diversification on the basis of majorization pre-ordering allows one to obtain comparisons of riskiness for portfolios of heavy-tailed and possibly dependent risks with arbitrary, rather than equal, weights.

Let us consider the portfolio VaR dealt with in Theorems 2.1.1 and 2.1.2 in the borderline case $\alpha = 1$ which corresponds to i.i.d. risks $X_1, \ldots, X_n$ with a symmetric Cauchy distribution $S_1(\sigma)$. As discussed in Sect. 2.1.2, these distributions are exactly at the dividing boundary between the class $\overline{\mathcal{CSLC}}$ in Theorem 2.1.1 and the class $\underline{\mathcal{CS}}$ in Theorem 2.1.2. Using (2.3) with $\alpha = 1$ we get that, for all $w = (w_1, \ldots, w_n) \in \mathcal{I}_n$, $Z_w = \sum_{i=1}^n w_i X_i =^d X_1$. Consequently, for all $q \in (0, 1)$, the value at risk $\mathrm{VaR}_q(Z_w) = \mathrm{VaR}_q(X_1)$ is independent of w and is the same for all portfolios of risks X_i with weights $w \in \mathcal{I}_n$, $i = 1, \ldots, n$. Thus, in such a case, diversification of a portfolio has no effect on riskiness of its return. Similarly, for general weights $w = (w_1, \ldots, w_n) \in \mathbf{R}_+^n$, property (2.3) with $\alpha = 1$ implies $Z_w = \sum_{i=1}^n w_i X_i =^d (\sum_{i=1}^n w_i) X_1$. Thus, the value at risk $\mathrm{VaR}_q(Z_w) = (\sum_{i=1}^n w_i) \mathrm{VaR}_q(X_1)$ is independent of w so long as $\sum_{i=1}^n w_i$ is fixed. Consequently, $\mathrm{VaR}_q(Z_w)$ is both Schur-convex (as in Theorem 2.1.1) and Schur-concave (as in Theorem 2.1.2) in $w \in \mathbf{R}_+^n$ for i.i.d. risks $X_i \sim S_\alpha(\sigma, 0, 0)$ that have symmetric Cauchy distributions with $\alpha = 1$ (see Marshall et al. 2011; Proschan 1965, p. 492, for similar properties of tail probabilities of Cauchy distributions).[6]

Remark 2.1.1 From Theorem 2.1.1 it follows that if X_1 and X_2 are i.i.d. risks such that $X_i \sim \overline{\mathcal{CSLC}}$, $i = 1, 2$, then $\mathrm{VaR}_q(X_1 + X_2) < \mathrm{VaR}_q(X_1) + \mathrm{VaR}_q(X_2)$ and $\mathrm{VaR}_q(\lambda X_1 + (1 - \lambda) X_2) < \lambda \mathrm{VaR}_q(X_1) + (1 - \lambda) \mathrm{VaR}_q(X_2)$ for all $q \in (0, 1/2)$ and any $\lambda \in (0, 1)$. That is, the VaR exhibits subadditivity and convexity, and is thus a coherent measure of risk for the class $\overline{\mathcal{CSLC}}$ (see Sect. 2.1.1 for the definition of coherent risk measures and coherency axioms in the case of the VaR). On the other hand, Theorem 2.1.2 implies that $\mathrm{VaR}_q(X_1) + \mathrm{VaR}_q(X_2) < \mathrm{VaR}_q(X_1 + X_2)$ and $\lambda \mathrm{VaR}_q(X_1) + (1 - \lambda) \mathrm{VaR}_q(X_2) < \mathrm{VaR}_q(\lambda X_1 + (1 - \lambda) X_2)$ for all $q \in (0, 1/2)$, $\lambda \in (0, 1)$ and i.i.d. risks $X_1, X_2 \sim \underline{\mathcal{CS}}$. Consequently, subadditivity and convexity are always violated for risks with extremely heavy-tailed distributions. In such a case, the VaR is not a coherent risk measure even in the case of independence which is "the worst case scenario" for diversification failure.

Remark 2.1.2 From the counterexamples constructed in Artzner et al. (1999), Embrechts et al. (2002) and Chap. 6 in McNeil et al. (2005) it follows that the VaR, in general, fails to satisfy the subadditivity and convexity properties. From the analysis similar to Examples 6 and 7 in Embrechts et al. (2002) and Chap. 12 in Bouchaud and Potters (2004) it follows that subadditivity of the VaR holds for distributions with power-law tails (1.3) and sufficiently small values of the loss probability q if $\zeta > 1$. Subadditivity is violated for power-law distributions (1.3) and sufficiently small values of the loss probability q if $\zeta < 1$. More generally, let, similar to Example 7 in Embrechts et al. (2002) and Sect. 12.1.2 in Bouchaud and Potters (2004), X_1 and X_2 be two i.i.d. risks with regularly varying heavy tails: $P(X_1 < -x) = L(x)/x^\zeta$, $\zeta > 0$, as $x \to +\infty$, where and $L(x)$ is a

---

[6]From the proof of Theorems 2.1.1 and 2.1.2 and this property it follows that the theorems continue to hold for convolutions of distributions from the classes $\overline{\mathcal{CSLC}}$ and $\underline{\mathcal{CS}}$ with Cauchy distributions $S_1(\sigma)$.

slowly varying at infinity function, that is $L(\lambda x)/L(x) \to 1$, as $x \to +\infty$, for all $\lambda > 0$ (see Embrechts et al. 1997; Zolotarev 1986, p. 8). Using the property that $\lim_{x \to +\infty} P(X_1 + X_2 < -x)/P(X_1 < -x/2^{1/\zeta}) = 1$ (see Lemma 1.3.1 in Embrechts et al. 1997 and Sect. 12.1.2 in Bouchaud and Potters 2004), one gets that $\lim_{q \to 0} \text{VaR}_q(X_1 + X_2)/(\text{VaR}_q(X_1) + \text{VaR}_q(X_2)) = 2^{1/\zeta - 1}$. Consequently, the subadditivity property holds for the VaR *asymptotically* as $q \to 0$ if $\zeta > 1$ and is violated as $q \to 0$ if $\zeta < 1$ (the important paper by Embrechts et al. (2009), shows that the above conclusions on the asymptotic, as $q \to 0$, subadditivity properties of the VaR for $\zeta > 1$ and their violations for $\zeta < 1$ continue to hold and are the same as in the case of independence for all risks with general Archimedean copula dependence structure that includes many models with contagion effects).[7] The implications of Theorems 2.1.1 and 2.1.2 for the VaR coherency in Remark 2.1.1 are qualitatively different from the counterexamples available in the literature and the above asymptotic considerations. This is because the VaR comparisons in Remark 2.1.1 hold *regardless* of the value of q and are valid for the *whole* wide classes of heavy-tailed risks. From the results in Sect. 5 in Ibragimov (2009b) it follows that similar VaR comparisons and conclusions, with arbitrary q's, also hold for a wide class of heavy-tailed dependent risks affected by common shocks.

Remark 2.1.3 Theorems 2.1.1 and 2.1.2 imply corresponding results on majorization properties of the tail probabilities $\xi(w, x) = P(\sum_{i=1}^{n} w_i X_i > x)$, $x > 0$, of linear combinations of heavy-tailed r.v.'s $X_1, \ldots, X_n$. These properties generalize the results in the seminal work by Proschan (1965) who showed that the tail probabilities $\xi(w, x)$ are Schur-convex in $w = (w_1, \ldots, w_n) \in \mathbf{R}^n_+$ for all $x > 0$ for i.i.d. r.v.'s $X_i \sim \mathcal{LC}, i = 1, \ldots, n$.[8,9] Schur-convexity of $\xi(w, x)$ for $X_i \sim \mathcal{LC}$ implies that the VaR comparisons in Theorem 2.1.1 hold for i.i.d. log-concavely distributed risks. The results in Proschan (1965) have been applied to the analysis of many problems in statistics, econometrics, economic theory, mathematical evolutionary theory, and other fields. One should note here that applicability of these majorization results and their analogs for other classes of distributions to portfolio VaR theory has not been recognized in the previous literature even in the case of i.i.d. log-concavely distributed risks.

A number of papers in probability and statistics have focused on extension of Proschan's results (see, among others, Chan et al. 1989; Jensen 1997; Ma 1998 and

---

[7]See also Ibragimov et al. (2014) for the analysis of the interplay of dependence modeled using different copula structures, the degree of heavy-tailedness and the values of loss probabilities and disaster levels in problems of portfolio diversification in VaR frameworks.

[8]Proschan's (1965) results are an example of many results and inequalities in probability whose formulation or proofs are based on majorization theory (for further examples of such inequalities, see Chap. 3 in Marshall et al. 2011).

[9]The main results in Proschan (1965) are reviewed in Sect. 12.J in Marshall et al. (2011). The work by Proschan (1965) is also presented, in a rearranged form, in Sect. 11 of Chap. 7 in Karlin (1968). Peakedness results in Proschan (1965) and Karlin (1968) are formulated for "PF2 densities," which is the same as "log-concave densities."

the review in Tong 1994). However, in all the studies that dealt with generalizations of the results, the majorization properties of the tail probabilities were of the same type as in Proschan (1965). Namely, the results gave extensions of Proschan's results concerning *Schur-convexity* of the tail probabilities $\xi(a, x)$, $x > 0$, to classes of r.v.'s more general than those considered in Proschan (1965). Analogues of Theorems 2.1.1 and 2.1.2 for the tail probabilities $\xi(a, x)$, on the other hand, provide the first general results concerning *Schur-concavity* of $\xi(a, x)$, $x > 0$, for certain wide classes of r.v.'s. According to these results, the class of distributions for which Schur-convexity of the tail probabilities $\xi(a, x)$ is replaced by their Schur-concavity is precisely the class of distributions with extremely heavy-tailed densities.[10]

2.1.4 Implications for Econometric and Statistical Methods[11]

Similar to the portfolio VaR analysis, heavy-tailedness presents a challenge for applications of standard statistical and econometric methods. In particular, as pointed out by Granger and Orr (1972) and in a number of more recent studies (see, among others, Chap. 7 in Cont 2001; Davis and Mikosch 1998; Embrechts et al. 1997; Mikosch and Stărică 2000, and the references therein) many classical approaches to inference based on variances and (auto)correlations such as regression and spectral analysis, least squares methods and autoregressive models may not apply directly in the case of heavy-tailed observations with infinite second or higher moments.

An important simple illustration is provided by the failure of the Law of Large Numbers (LLN) for observations with infinite first moments and variances. When more information about the structure of heavy-tailedness is available, one can obtain more refined results that point out to crucial differences between inference in moderately heavy-tailed and extremely heavy-tailed populations.

---

[10]The analysis of tail probabilities of linear combinations of r.v.'s is related to the field of probability and moment inequalities in probability and statistics (see, among others, Sect. 12 in de la Peña and Giné 1999; de la Peña et al. 2003; Hansen 2015; Ibragimov and Ibragimov 2008; Ibragimov and Sharakhmetov 1997, 2002; Marshall et al. 2011; Nze and Doukhan 2004; Utev 1985, and the references therein for a number of results in the field and their statistical and econometric applications).

[11]This section draws upon material from the following articles:

Ibragimov (2007) "Efficiency of linear estimators under heavy-tailedness: Convolutions of α-symmetric distributions," *Econometric Theory*, Volume 23(3), pp. 501–517 (2010) © Cambridge University Press, reproduced with permission, and

Ibragimov (2009a) "Heavy-tailed densities," "The New Palgrave Dictionary of Economics," Eds. Steven N. Durlauf and Lawrence E. Blume, Palgrave Macmillan, reproduced with permission of Palgrave Macmillan. The full published version of this publication is available from: http://www.dictionaryofeconomics.com/article?id=pde2009_H000191.

Consider the problem of estimating the location parameter μ in the model

$$X_i = \mu + \eta_i, \tag{2.5}$$

where η_i are i.i.d. errors with an absolutely continuous symmetric distribution. Given a random sample $X_1, \ldots, X_n$ that follows (2.5) with center μ, and weights $w = (w_1, \ldots, w_n) \in \mathbf{R}_+^n$, denote by $\hat{\theta}_n(w)$ the linear estimator $\hat{\theta}_n(w) = \sum_{i=1}^n w_i X_i$ and by $\psi(w, \epsilon)$, $\epsilon > 0$, its tail probability $\psi(w, \epsilon) = P(|\hat{\theta}_n(w) - \mu| > \epsilon)$. As before, we also denote $\mathcal{I}_n = \{w = (w_1, \ldots, w_n) \in \mathbf{R}_+^n : \sum_{i=1}^n w_i = 1\}$.

It is well-known that, if $E\eta_i^2 < \infty$, then the sample mean $\overline{X}_n = (1/n) \sum_{i=1}^n X_i$ is the best linear unbiased estimator (BLUE) of the population mean $\mu = EX_i$. That is, $\overline{X}_n$ is the most efficient estimator of μ among all unbiased linear estimators $\hat{\theta}_n(w)$ in the sense of variance comparisons: $Var(\overline{X}_n) \leq Var(\hat{\theta}_n(w))$ for all $w \in \mathcal{I}_n$.

The definition of efficiency based on variance breaks down in the case of heavy-tailed populations with infinite second moments. A natural approach to comparison of performance of estimators under heavy-tailedness is to order them by likelihood of observing large deviations from the population parameter of interest being estimated. This approach relies on the concept of peakedness of r.v.'s and leads to the following definition.

Let $\hat{\theta}(v)$ and $\hat{\theta}(w)$ be two linear estimators of the parameter μ in model (2.5). The estimator $\hat{\theta}(v)$ is said to be more efficient than $\hat{\theta}(w)$ in the sense of peakedness (P-more efficient than $\hat{\theta}(w)$ for short) if $P(|\hat{\theta}(v) - \mu| > \epsilon) < P(|\hat{\theta}(w) - \mu| > \epsilon)$ for all $\epsilon > 0$. The property of being P-less efficient is defined in a similar way. Roughly speaking, $\hat{\theta}^{(1)}$ is P-more efficient than $\hat{\theta}^{(2)}$ if the distribution of $\hat{\theta}^{(1)}$ is more concentrated about the true parameter μ than is that of $\hat{\theta}^{(2)}$.

In view of Remark 2.1.3, the results on portfolio VaR comparisons in Sect. 2.1.3 can be equivalently formulated in terms of efficiency comparisons for linear estimators of a location parameter in heavy-tailed models (2.5). Namely, if the errors η_i in model (2.5) are moderately heavy-tailed: $\eta_i \sim \overline{CSLC}$, then, for $v, w \in \mathcal{I}_n$, the linear estimator $\hat{\theta}_n(v)$ is P-more efficient than $\hat{\theta}_n(w)$ if $v \prec w$ and v is not a permutation of w (equivalently, $\psi(v, \epsilon)$ is strictly Schur-convex in $v = (v_1, \ldots, v_n) \in \mathbf{R}_+^n$ for all $\epsilon > 0$). Also, in this case, the sample mean $\overline{X}_n$ is the BLUE of the location parameter (the population mean) μ in the sense of P-efficiency: the sample mean is P-more efficient than any other linear unbiased estimator $\hat{\theta}_n(w) = \sum_{i=1}^n w_i X_i$, $w \in \mathcal{I}_n$. In particular, $\overline{X}_n$ exhibits monotone consistency for μ in the sense that $P(|\overline{X}_n - \mu| > \epsilon)$ converges to zero strictly monotonically in n for all $\epsilon > 0$.

Similar to the effects of diversification in portfolio VaR models, the above efficiency properties of linear estimators are reversed in the case of extremely heavy-tailed models (2.5) with $\eta_i \sim \mathcal{CS}$. In the extremely heavy-tailed case, for $v, w \in \mathcal{I}_n$, the linear estimator $\hat{\theta}_n(v)$ is P-less efficient than $\hat{\theta}_n(w)$ if $v \prec w$ and v is not a permutation of w (equivalently, the function $\psi(v, \epsilon)$ is strictly Schur-concave in $v = (v_1, \ldots, v_n) \in \mathbf{R}_+^n$ for all $\epsilon > 0$). Further, under extreme heavy-tailedness in model (2.5) with $\eta_i \sim \mathcal{CS}$, the sample mean $\overline{X}_n = (1/n) \sum_{i=1}^n X_i$ is P-less efficient than any other linear estimator $\hat{\theta}_n(w) = \sum_{i=1}^{n} w_i X_i$ with $w \in \mathcal{I}_n$. In particular, P-efficiency of the sample mean $\overline{X}_n$ decreases with n, that is, $P(|\overline{X}_{n+1} - \mu| > \epsilon) > P(|\overline{X}_n - \mu| > \epsilon) > P(|X_1 - \mu| > \epsilon)$ for all $n \geq 1$ and all $\epsilon > 0$.

The results and conclusions in the next sections show that, similar to the portfolio VaR analysis and the efficiency properties of linear estimators, many models in economics and related fields are robust to heavy-tailedness assumptions as long as the distributions entering these assumptions are moderately heavy-tailed. However, the implications of these models are reversed for distributions with sufficiently heavy-tailed densities.

2.2 Diversification Analysis: Bounded Case[12]

Let, as in Sect. 1.3, $I(\cdot)$ stand for the indicator function. For an r.v. (risk) X, we define its a-truncated version by $Y(a) = X$ for $|X| \leq a$, $Y(a) = -a$ for $X < -a$ and $Y(a) = a$ for $X > a$. In other words, $Y(a) = a \cdot sign(X_i) + XI(|X| \leq a)$, where $sign(x)$ is the sign of x defined by $sign(x) = 1$ if $x > 0$, $sign(0) = 0$ and $sign(x) = -1$ otherwise.[13] We will also use the notation X^a instead of $Y(a)$ for the a-truncated version of X.

Let $0 \leq r < 1$. Following the framework of Roy (1952) safety-first, given a random loss (risk) Z, we are interested in analyzing the probability $P(Z < -z)$ of getting losses greater than a certain disaster level $z > 0$. Furthermore, as before, given a loss probability $q \in (0, 1)$ and an r.v. (risk) Z, we denote by $\text{VaR}_q(Z)$ the VaR of Z at level q, that is, the negative of its q-quantile.[14]

Throughout this section, $X_1, X_2, \ldots$ is a sequence of i.i.d. risks with distributions from the class $\mathcal{CS}$. For $a > 0$, denote by $Y_i(a)$ the a-truncated versions of X_i's, as defined above. Let, as in Sect. 2.1, $\mathcal{I}_n = \{w = (w_1, \ldots, w_n) \in \mathbf{R}_+^n : \sum_{i=1}^n w_i = 1\}$. For $a > 0$ and $w \in \mathcal{I}_n$, denote by $Y_w(a)$ the risk of the portfolio of bounded risks $Y_1(a), \ldots, Y_n(a)$ with weights w: $Y_w(a) = \sum_{i=1}^{n} w_i Y_i(a)$. Evidently, the risk $Y_{\tilde{w}_n}(a)$ of the portfolio of $Y_1(a), \ldots, Y_n(a)$ with equal weights $\tilde{w}_n = \left(\frac{1}{n}, \frac{1}{n}, \ldots, \frac{1}{n}\right)$ is given by the sample mean of $Y_i(a)'$s: $Y_{\tilde{w}_n}(a) = \frac{1}{n} \sum_{i=1}^{n} Y_i(a)$.

---

[12]This section is based on the article Ibragimov and Walden (2007), which was published in the *Journal of Banking and Finance*, Vol. 31, Issue 8, pp. 2551–2569, Copyright Elsevier (2007).

[13]This definition of truncation moves probability mass to the edges of the distributions. The results in this section continue to hold for the more commonly used truncations $XI(|X| \leq a)$ which move probability mass to the center.

[14]That is, in the case of an absolutely continuous risk Z, $P(Z \leq \text{VaR}_q(Z)) = q$.

The problem faced by a holder of bounded risks $Y_1(a), \ldots, Y_n(a)$ consists in minimizing the disaster probability $P\left(\sum_{i=1}^{n} w_i Y_i(a) < -z\right)$ over the portfolio weights $w \in \mathcal{I}_n$. Let, as in Sect. 2.1.1, $w_{[1]} \geq \ldots \geq w_{[n]}$ denote the components of $w \in \mathcal{I}_n$ in decreasing order. Obviously, $w_{[1]} = 1$ implies that w is a permutation of the vector $(1, 0, \ldots, 0)$. In such a case, evidently, the portfolio with weights w consists of only one risk, and, thus, $Y_{\tilde{w}_n}(a)$ has the same distribution as $Y_1(a)$. In addition, for $w \in \mathcal{I}_n$, let $(w^{(1)}, w^{(2)}) = \left(max[0.5, w_{[1]}], min[0.5, 1 - w_{[1]}]\right)$.

Theorem 2.1.2 (see also Remark 2.1.3) implies that the stylized facts that portfolio diversification is always preferable are violated for a wide class of *extremely heavy-tailed* risks $X_i \sim \mathcal{CS}$ with *unbounded* support. In such a setting, diversification of a portfolio of the risks increases the probability of going over a given disaster level.

We now expand the analysis in Sect. 2.1.3 to risks with bounded support. A summary of the results we will provide is given in Fig. 2.1. The situation with thin-tailed and moderately heavy-tailed i.i.d. risks (see the Introduction and Theorem 2.1.1) is according to line A in the figure: Diversification is always to be preferred, regardless of the number of risks. The opposite situation is in line C for extremely heavy-tailed risks, where diversification will never be preferred (Theorem 2.1.2). The intermediate case is line B for settings like bounded risks considered in this section where diversification is suboptimal up to a certain number of risks (similar to C), but becomes preferable when enough assets are available (similar to A).

The following theorem is the analogue of Theorem 2.1.2 and its implications for tail probabilities (see Remark 2.1.3) in the case of *bounded* risks. The theorem shows that diversification continues to be disadvantageous for truncated extremely heavy-tailed distributions, given a VaR-based definition of riskiness. The results show, in particular, that for any number $n \geq 2$ and any given disaster level $z > 0$,

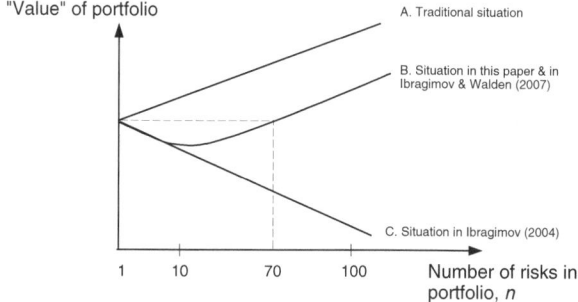

Fig. 2.1 Value of diversification. (*A*) Thin-tailed and moderately heavy-tailed risks: the value increases monotonically and it is always preferable to add another risk to portfolio (Theorem 2.1.1). (*B*) Bounded truncations of heavy-tailed distributions: up to a certain number of assets, value decreases with diversification (Theorem 2.2.1). (*C*) Extremely heavy-tailed risks: value always decreases with diversification (Theorem 2.1.2)

there exist n risks with *finite* support with the property that a diversified portfolio is riskier than a portfolio consisting of only one risk.

Theorem 2.2.1 *Let $n \geq 2$ and let $w \in \mathcal{I}_n$ be a portfolio of weights with $w_{[1]} \neq 1$. For any $z > 0$ and all sufficiently large $a > A$, the following inequality holds:*

$$P\Big(Y_w(a) < -z\Big) > P\Big(Y_1(a) < -z\Big).$$

Remark 2.2.1 The threshold value A in the length of the support of bounded risks with suboptimal diversification in Theorem 2.2.1 depends on distributional properties (and the degree of heavy-tailedness) of risks X_i, their number n, the portfolio weights, w and the disaster level z. Ibragimov and Walden (2007) provide the analysis of values of the threshold levels A for different heavy-tailed risks X_i.

Remark 2.2.2 Theorem 2.2.1 shows that, for a *specific* loss probability $q \in (0, 1/2)$, there exists a sufficiently large value of the length of the distributional support a such that the VaR $\mathrm{VaR}_q\big(Y_w(a)\big)$ of $Y_w(a)$ at level q is greater than the value at risk $\mathrm{VaR}_q\big(Y_1(a)\big)$ of $Y_1(a)$ at the same level: $\mathrm{VaR}_q\big(Y_w(a)\big) > \mathrm{VaR}_q(Y_1(a))$. This inequality between the risks $Y_w(a)$ and $Y_1(a)$ holds for the particular *fixed* loss probability q. In the comparisons of the values at risk $\mathrm{VaR}_q\big(Y_w(a)\big)$ and $\mathrm{VaR}_q(Y_1(a))$, the length of the interval needed for the reversals of the stylized facts on the portfolio diversification depends on q. This is a crucial difference compared with Theorem 2.1.2 and Remark 2.1.3, where the inequalities hold for all disaster level $z > 0$ and all loss probabilities $q \in (0, 1/2)$.

2.3 Insurance Markets: Non-diversification Traps[15]

Catastrophe insurance provides compensation for losses created by such natural risks as earthquakes, floods, and wind damage, as well as man-created risks including terrorism. Over the past 15 years, most private-market property and casualty (P&C) insurance firms stopped offering coverage against catastrophe risks, usually in the wake of a major event. Key private market failures include Florida hurricane insurance after Hurricane Andrew in 1992, California earthquake insurance after the Northridge quake of 1994, and, most recently, U.S. terrorism insurance after the 9/11 attack. Catastrophe insurance markets have also failed worldwide in most developed countries. In response, state and federal governments have been forced to try to replace or to revive the private markets, the most recent example being the U.S.

---

[15]This section is based on the article Ibragimov et al. (2009), which was published in the *Review of Financial Studies*, Vol. 22, Issue 3, pp. 959–993.

Terrorism Risk Insurance Act of 2002 (TRIA) and its 2005 extension. Unfortunately, the government interventions are generally considered to be quite inefficient.[16]

Understanding the source of the private market failures is essential if more effective remedies are to be found. A fundamental question is why catastrophe risks are "uninsurable" for the private insurance firms. Asymmetric information—adverse selection and/or moral hazard—is the common textbook explanation for insurance market failures, but there seems to be little role for asymmetric information with respect to natural disasters or terrorism attacks.[17] Imprecision in estimating the underlying stochastic process is also sometimes suggested as a basis for "uninsurability," but even if parameter imprecision raises the cost of insurance, perhaps due to ambiguity aversion, it is unclear why it would cause the market to fail; see Froot (2001).

A third basis for "uninsurability" is that the possible losses may exceed the capital resources of the P&C insurance industry; see Cummins et al. (2002). For example, the losses created by war or by terrorist use of weapons of mass destruction (WMD) could readily exceed the capital resources of all P&C firms. The deadweight costs of bankruptcy, including a loss in the value of the managers' human capital, could then motivate the unwillingness of firms to participate in the catastrophe insurance markets. War and WMD risks, however, have long been excluded from most insurance contracts, without jeopardizing the availability of coverage for standard risks.

War and WMD risks aside, size alone does not appear to explain the recent failure of so many different catastrophe insurance markets. Table 2.1 shows the ten most costly insurance losses since 1970 as compiled by SwissRe (2006). The losses created by the Katrina hurricane of 2005 were USD 45 billion, followed by the Florida Hurricane Andrew of 1992 (USD 22 billion), and the 9/11 terrorist attack (USD 21 billion). In comparison, the capital resources of the P&C insurance industry at year-end 2005 totaled approximately USD 446 billion, and in each year since 2001, the P&C industry has *increased* its capital resources (net of losses) by at least USD 29 billion.[18]

Although P&C industry resources can cover most catastrophic risks, coverage is provided at the level of individual firms, not the industry. Furthermore, insurers tend to specialize in geographic regions and particular lines of coverage, putting individual firms at potentially high risk to a specific catastrophic event. Regulation is a major cause of the geographic and insurance line specialization, since U.S.

---

[16]See Cummins (2006), Jaffee (2006a), and Jaffee and Russell (2006) for recent discussions and references to the literature. OECD (2005a) and OECD (2005b) discuss government interventions around the world to reactivate terrorism insurance. Kunreuther and Michel-Kerjan (2006) discuss the specific issue of terrorism insurance in the United States.

[17]There is generally open access to scientific forecasts of natural disasters, much of it provided by governments. Terrorists may be more strategic in their choice of targets, but this does not create a moral hazard on the part of those purchasing terrorism insurance (unless the terrorists particularly target insured properties).

[18]These data are from the Insurance Information Institute; see http://iii.org/media/industry/.

Table 2.1 The world's ten costliest insurance events

Date	Event	Insured losses[a] (USD Billion)
2005, August 24	Hurricane Katrina	45
1992, September 23	Hurricane Andrew	22
2001, September 11	World Trade Center	21
2004, January 17	Northridge Earthquake	18
2004, September 2	Hurricane Ivan	12
2005, September 20	Hurricane Rita	10
2005, October 16	Hurricane Wilma	10
2004, August 11	Hurricane Charley	8
1991, September 27	Typhoon Mireille (Japan)	8
1990, January 25	Storm Daria (Europe)	7

Source: SwissRe (2006)
[a]Property and Business interruption; excludes liability and life insurance

insurance firms must be chartered separately in each state in which they operate and they face substantial fixed costs for marketing and for developing actuarial expertise for each line and for each state.[19] The result is that relatively few catastrophe insurance firms operate in each state and for each catastrophe line. Risk-averse executives with ties to their own firm may wish to avoid such an undiversified position.

Reinsurance firms exist, of course, precisely to redistribute risks, allowing individual insurers to match their retained risks with their capital resources. Thus, if reinsurance markets function efficiently, then capital adequacy at the industry level is in fact the relevant measure. Unfortunately, the proximate cause of the observed failures of the primary catastrophe insurance markets has been precisely the failure of the associated reinsurance markets. For this reason, the fundamental question is why the reinsurance markets for catastrophe risks have largely failed.

In this section, we argue that the observed dynamic pattern of widely varying supply conditions for catastrophe insurance and reinsurance could reflect a multiple equilibrium system, with the market sometimes reaching a coordinated reinsurance/diversification equilibrium, but at other times falling into what we call a nondiversification trap. The term is related to poverty traps and development

---

[19]Insurance is unique among U.S. financial services in that it is regulated in the United States *only* at the state level. The structure of a catastrophe insurance market is well illustrated by California's earthquake risk market. As of 2005, 70 % of the coverage was provided by the California Earthquake Authority, an entity created by the State of California following the 1994 Northridge quake. With no major quakes since then, private insurers have slowly reentered the market, now representing about 30 % of the market. However, still only 35 private insurance groups are offering California earthquake coverage (based on annual written premiums of USD 1 million or more). Furthermore, the top 5, 10, and 20 firms represent 46, 66, and 89 % of the total private market, respectively.

traps in economic growth theory (Azariadis and Stachurski 2006; Barro and Sala-i-Martin 2004). It denotes a situation where there are two possible equilibria: a diversification equilibrium in which insurance is offered and there is full risk sharing through the reinsurance market, and a nondiversification equilibrium, in which the reinsurance market is not used, and no insurance is offered at all. A move from the nondiversification equilibrium to the diversification equilibrium has to be coordinated by a large number of insurers and reinsurers, which may be difficult to achieve through a market mechanism. Therefore, there may be a role for a centralized agency to ensure that the diversification equilibrium is reached—for example, by mandating that insurance must be offered (as in the case of the U.S. Terrorist Risk Insurance Act of 2002 and in the corresponding government plans in most European countries).

Consequently, our discussion and model focus on reinsurance as the mechanism that could be used to coordinate the diversification equilibrium. A functioning reinsurance industry, however, requires that the primary insurers be willing to write policies in anticipation that other insurers will do the same and that the reinsurers will pool all the risks, to reach the global diversification outcome. Our model will determine the conditions under which such an equilibrium can and cannot occur.

The existence of nondiversification traps depends crucially on there being conditions under which diversification becomes suboptimal for the individual insurers. This is contrary to the traditional framework in which diversification is always preferred (see, e.g., Samuelson 1967b for equity investments and Froot and Posner 2002 for insurance). The traditional framework uses concave optimization (e.g., via expected utility), with thin-tailed risks (e.g., normal distributions), and without distortions (unlimited liability, no frictions and no fixed costs). If any of these assumptions fails, diversification may not always be preferred. Our discussion focuses, in particular, on the impact of heavy left-tailed distributions (implying a nonnegligible probability for large negative outcomes) as the defining property of catastrophic risks.

Figure 2.1 in the previous section provides an intuition for how nondiversification traps can arise for insurance. Consider a situation in which there is a maximum number of distinct risks that an individual insurance provider can take on—e.g. $N = 10$. The constraint of the maximum number of risks that an individual firm can accept is in line with our regulatory discussion earlier and can also be motivated by capacity constraints, capital requirements, and segmented markets. The three lines, A–C, in the figure describe the value (e.g., measured as a certainty equivalent) of holding a diversified portfolio of n risks as a function of n. In this section we will study a model in which the value is a U-shaped function of the number of risks, corresponding to line B (also studied in the previous section). In this case, for any individual insurance provider, diversification will clearly be suboptimal as the value decreases in n for $n \leq N = 10$. However, if there are M insurance providers in the market, they could potentially meet in a reinsurance market, pool the risks, and reach full diversification with NM risks. For this to be preferred to nondiversification, at least $M = 7$ insurance providers must pool the risks. This is a very different situation compared with the traditional situation in line A, in which each individual insurance provider will choose maximal diversification into N risks, and in which two insurance providers can always improve their situation by pooling their risks in a reinsurance market. For line B, there may be a coordination problem.

2.3.1 Risk Pooling

We begin by studying the potential value of risk sharing among multiple risk-takers. We first develop the intuition and then, in the following sections, prove the results rigorously for a model of a reinsurance market.

We study the behavior of risk-takers. Because we focus on the context of risk-taking insurance companies, we will refer to these risk-takers as insurers. We assume that the number of insurers is bounded by M and that all insurers are expected utility optimizers with identical strictly concave utility functions, u.

We assume that there is limited liability. Clearly, real-world insurance firms have limited liability and may default in some states of the world. This case is increasingly studied in the insurance literature; see Cummins et al. (2002), Cummins and Mahul (2003), and Mahul and Wright (2004). For catastrophe insurance, with heavy-tailed distributions, there is an effectively nonzero (although small) probability that such a catastrophic event will create default. Technically, limited liability is needed in the model, because with heavy-tailed distributions, the expected payoffs and values are not otherwise defined. We shall, however, see that the probability for default is small in equilibrium. Moreover, we will show that our results are not driven by the convexity of payoffs introduced by limited liability: For markets with large aggregate risk-bearing capacity, our results will apply only if distributions are heavy-tailed. The assumption of limited liability is modeled by insurers being liable to cover losses only up to a certain amount, k. If losses exceed k, an insurer pays k, but defaults on any additional loss.[20] Thus, if an insurer takes on a random risk of X, the effective outcome for the insurer once X is realized is

$$V(X) = \begin{cases} X, & \text{if } X \geq -k, \\ -k, & \text{if } X < -k. \end{cases} \tag{2.6}$$

In the special case when there is no limited liability—i.e., when $k = \infty$—we have $V(X) = X$ for all X. If $k < \infty$, u needs only to be defined on $[-k, \infty)$ and we can without loss of generality assume that $u(-k) = 0$.

---

[20] We assume that a third party, perhaps the government, covers the excess losses to policy holders. This avoids the complications of any impact on policyholder demand.

Assuming i.i.d. risks $X_1, X_2, \ldots$, we wish to study the expected utility of s agents, who share j risks equally. We therefore define the random variable $z_{j,s} = (\sum_{i=1}^{j} X_i)/s$, with cdf $F_{j,s}$. The expected utility of such risk sharing is:

$$U_{j,s} \stackrel{\text{def}}{=} Eu(V(z_{j,s})) = \int_{-k}^{\infty} u(x)dF_{j,s}(x). \tag{2.7}$$

Firms are usually considered to be risk neutral. However, an expected utility setup with concave utility can effectively arise if there are financial imperfections as, for example, assumed in Froot et al. (1993). If such financial imperfections are present, the value of the firm will be given by a concave transformation of the payoffs, which is effectively identical to our expected utility setup. Another motivation for risk-averse firm behavior is that executives with major financial and human capital investments in their own firm wish to avoid risky positions.

Insurers face two constraints, one that limits the aggregate amount of their risks, and the other that limits the size of individual risks. The aggregate limitation is driven, for example, by capital requirements. This aggregate limit is imposed by assuming that each insurer can bring at most N risks "to the table." Thus, we have $1 \le s \le M$, $1 \le j \le Ns$. The second constraint is that each risk, X_i, is indivisible, so it cannot be split between insurers in a primary insurance market. As discussed in the introduction, real-world catastrophe insurance markets are segmented in this way because relatively few insurers operate in each state and in each line.

When returns are independently normally distributed, it is well known that one can always add an asset to a portfolio and strictly increase the agent's utility via the appropriate selection of weights. In this case, $U_{j,s}$ is strictly increasing in s for each j (an immediate consequence of Samuelson 1967b). In this situation we can expect a reinsurance market to work well and insurance to be offered for a maximal number of risks, NM. The argument is based on the fact that each insurer will choose to diversify fully, regardless of what the other $M - 1$ insurers do. We denote this the *traditional situation*.

The situation is very different when we have limited liability and heavy-tailed distributions. We consider i.i.d. Bernoulli–Cauchy distributed risks, $\tilde{X}_i$—i.e.,

$$\tilde{X}_i = \begin{cases} \lambda, & \text{with probability } 1 - q, \\ X_i, & \text{with probability } q, \end{cases}$$

where $X_i = \mu + \xi_i$, $\xi_i \sim S_1(\sigma)$, are i.i.d. Cauchy r.v.'s with the location parameter μ, the scale parameter σ and the density (2.1). In other words, the r.v.'s X_i are "mixtures" of degenerate and Cauchy r.v.'s. Clearly, the risks $\tilde{X}_i$ can be written as

$$\tilde{X}_i = \lambda(1 - \epsilon_i) + X_i \epsilon_i = \lambda + (\mu - \lambda)\epsilon_i + \sigma Y_i \epsilon_i, \tag{2.8}$$

where ϵ_i are i.i.d. nonnegative Bernoulli r.v.'s with $P(\epsilon_i = 0) = 1-q, P(\epsilon_i = 1) = q$ and $Y_i \in S_{0,1}$ are i.i.d. symmetric Cauchy r.v.'s with scale parameter $\sigma = 1$ that are independent of ϵ_i's.

For the above distributions, we say that $\tilde{X}_i \in \tilde{S}^q_{\lambda,\mu,\sigma}$. Here, λ can be thought of as the premium an insurance provider collects to insure against events that occur with probability q. For $q \ll 1$, this distribution is qualitatively similar to distributions for catastrophic risks: There is a small probability for a catastrophe to occur. However, if it does occur, the loss may be very large due to the heavy left tail of the Cauchy distribution. We use the Cauchy distribution for its analytical tractability (even though it, similar to the normal distribution—used, e.g., in Cummins 2006—has a nonzero right tail, which does not have a meaningful interpretation for catastrophic events). We assume limited liability ($k < \infty$) and the power utility function $u(x) = (x + k)^\alpha$, $\alpha \in (0, 1)$. Clearly, under the above assumptions, the expected utility for Bernoulli–Cauchy risks always exists.

In Fig. 2.2, we show expected utility for different total numbers of projects, j, and numbers of agents involved in risk sharing, s, with parameters $k = 100, \sigma = 1$, $\lambda = 1$, $\mu = -9$, $N = 20$, $M = 5$, $\alpha = 0.0315$, and $q = 0.05$. There is a crucial difference compared with the traditional situation. For a moderate number of risks, there is no way to increase expected utility compared with staying away from risks altogether. An insurer has the option of not entering the market and must therefore earn a utility premium to be willing to take on risk (i.e., to offer insurance). No insurer will therefore choose to invest in risks that cannot be pooled. Moreover, if an insurer believes that no other insurer will pool risks, he will not take

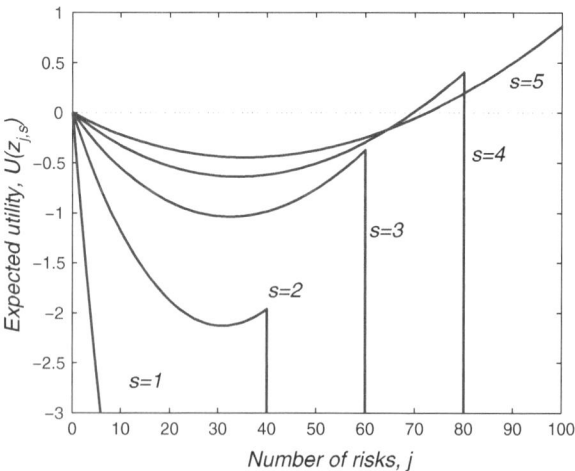

Fig. 2.2 Expected utility for insurers under different risk-sharing alternatives. s denotes the number of insurers sharing risks. j denotes the total number of risks. $U(z_{j,s})$ denotes the expected utility of an insurer as a function of j and s. Parameters: liability $k = 100$, total number of insurers $M = 5$, maximum number of risks per insurer $N = 20$, risk parameters: $\sigma = 1, \mu = -9, \lambda = 1$, $\alpha = 0.0315$ and $q = 5\%$ (see Eq. 2.8)

on risks, whether he can pool them or not. Thus, even though the situation with full diversification and risk sharing ($U_{NM,M}$) is preferred over the no-risk situation ($U_{0,1}$), at least four insurers must agree to pool risk for risk sharing to be worthwhile.

In this situation, there may be a coordination problem: Even though all agents would like to reach $U_{NM,M}$, they may be stuck in $U_{0,1}$. Clearly, the limited liability assumption is important: If liability were unlimited, no agent would ever take on risk. The situation would be as in Ibragimov (2005, 2009b), where diversification is always inferior. However, we note that the probability for default in the situation with full pooling and diversification is small: It is approximately 0.3 %.

The expected utility assumption is not crucial. Similar results would arise in a VaR framework—for example, with agents who trade off VaR versus expected returns for some risk level, α. The crucial property of the $U_{j,s}$ curves are that they are "U-shaped" in s. In Ibragimov and Walden (2007), it is shown that similar U-shaped curves occur as a function of diversification when the VaR measure is used. The specification in a VaR framework would be $U_{j,s} = F(\lambda, W)$, $\lambda = E(V(z_{j,s}))$, $W = \text{VaR}_\alpha(V(z_{j,s}))$, with $\partial F/\partial \lambda > 0$, and $\partial F/\partial W < 0$, and the analysis would be similar to the analysis we carry out in this section.

Our argument so far has been informal. We next make these diversification results rigorous by introducing a model of a reinsurance market where coordination plays a role—the *diversification game*. We will show that in the traditional situation, the only equilibrium is a diversification equilibrium, where *NM* risks are insured, whereas in the situation with heavy tails there is both a diversification equilibrium and a nondiversification equilibrium in which no insurance is offered.

2.3.2 A Reinsurance Market

We analyze a market in which insurance providers sell insurance against risks. For simplicity, we model the market in a symmetric setting: participants in reinsurance markets share risks equally. The setup is a two-stage game that captures the intuitive idea that insurance has to be offered before reinsurance can be pooled. The decision whether to offer primary insurance will be based on beliefs about how well-functioning (the future) reinsurance markets will be. If a critical number of participants is needed for reinsurance markets to take off, then nondiversification traps can occur. As we have already discussed the intuition behind nondiversification traps, Sects. 2.3.2, 2.3.3, and 2.3.4 focus on providing the theoretical foundation for the existence of nondiversification traps.

The two-stage *diversification game* describes the market. In the first stage, agents (insurance providers) simultaneously choose whether to offer insurance against a set of i.i.d. risks. In the second stage, the reinsurance market is formed and each agent chooses whether to participate or not. Agents who choose not to offer primary insurance are allowed to participate in the reinsurance market. Finally, all risks of agents participating in the reinsurance market are pooled, outcomes are realized and shared equally among participating agents.

2.3.2.1 Insurance Market

There are $M \geq 2$ agents (also referred to as insurance providers, insurance companies, or insurers). We use m, $1 \leq m \leq M$ to index these agents. There is a set of i.i.d. risks, $\mathcal{X}$, where each risk has cdf $F(x)$. Each agent chooses to take on a specific number of risks, $n_m \in \{0, 1, 2, \ldots, N\}$, where N denotes the maximum insurance capacity, forming a portfolio of risks $p_m \in \mathcal{P}_m$, where $p_m = \sum_{i=1}^{n_m} X_i$ and $X_i \in \mathcal{X}$. This is the first stage of the market. The risks are atomic (indivisible) and each risk can be chosen by at most one agent. We assume that there are enough risks available to exhaust capacity, i.e., $|\mathcal{X}| = NM$. Here, $|\mathcal{X}|$ denotes the cardinality of $\mathcal{X}$. As risks are i.i.d., only the distributional assumptions of the risks matter and we will not care about which insurance provider chooses which risk. The portfolio p_m is therefore completely characterized by the number of risks, n_m. The total number of risks insured is $\overline{N} = \sum_m n_m$.

Agents have liability to cover losses up to k, where $k \in (0, \infty]$. If losses exceed k for an agent, he defaults and pays k, and a third party, possibly the government, steps in and covers excess losses. The effective outcome under limited liability for agent m, taking on risk z_m, is therefore $V(z_m)$, where V is defined in (2.6). All agents have identical expected utility over risks, $U_m(z_m) = Eu(V(z_m))$, where u is defined and continuous on $[-k, \infty)$, is strictly concave, twice continuously differentiable on $(-k, \infty)$ and, if $k < \infty$, satisfies $u(-k) = 0$. The outcome of the first stage is summarized by $p = (p_1, \ldots, p_M) \in \mathcal{P} \overset{\text{def}}{=} \prod_{m=1}^{M} \mathcal{P}_m$.

2.3.2.2 Reinsurance Market

In the second stage of the game, named the *participation subgame*, the reinsurance market is formed. In this stage, agents have perfect knowledge about p. Each agent, $1 \leq m \leq M$, sequentially decides whether to participate in the market or not, as follows: First, agent 1 decides whether to participate. This is represented by the binary variable $q_1 \in \{0, 1\}$, where $q_1 = 1$ denotes that agent 1 participates in the reinsurance market and $q_1 = 0$ otherwise. Then, agent 2 decides whether to participate, observing agent 1's decision, etc. This is repeated until all M agents have decided. Previous agents' decisions are observable. If an agent is indifferent between participating and not participating, he will not participate. Agents who offer insurance, and participate, pool all their insurance in the reinsurance market—i.e., $q_m p_m$ is supplied to the reinsurance market by agent m. The total pooled risk is therefore $P = \sum_m q_m p_m$ and the number of risks is $R = \sum_m q_m n_m \in \{0, \ldots, NM\}$.

As noted, the two stages separate the choice of offering insurance from the creation of a reinsurance market, which can occur only when the risks are already insured. The total number of participating agents in the reinsurance market is $t = \sum_m q_m$. Finally, the pooled risks are split equally among agents participating in the reinsurance market—i.e., each participating agent receives a fraction $1/t$ of the pooled portfolio, P, with R risks.

The outcome of the participation subgame is summarized by $q = (q_1, q_2, \ldots, q_M) \in \{0, 1\}^M$ and the outcome of the total diversification game is thus completely characterized by (p, q). Moreover, the quintuple $\mathcal{G} = (u, F, k, N, M)$ completely characterizes the diversification game.

We study equilibrium outcomes (p, q) of a diversification game $\mathcal{G}$. As the second stage of the market is an M-step sequential game with perfect information, it is straightforward to calculate the unique subgame perfect equilibrium by backward induction (existence and uniqueness being guaranteed by Zermelo's theorem and by imposing the assumption that indifferent agents do not participate). A detailed setup for the participation subgame is given in Ibragimov et al. (2009). The equilibrium mapping of the participation game, for a specific first-stage realization, p, is a vector $q = \mathcal{E}(p) \in \{0, 1\}^M$. We use this mapping to simplify the analysis of the first stage of the diversification game. Specifically, in the first stage, all agents agree on $q = \mathcal{E}(p)$ as the outcome of the participation subgame, and therefore use it directly in their value function. This reduces the size of the strategy space considerably, while not having any effect on the (subgame perfect) equilibrium outcome. The sequence of events is shown in Fig. 2.3.

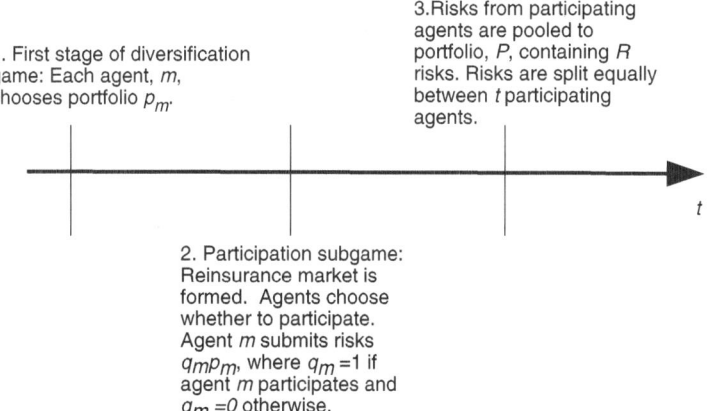

Fig. 2.3 Sequence of events: *1* Agents simultaneously choose risk portfolio, p_m. *2* Reinsurance pool $P = \sum_m q_m p_m$ is formed. Agents sequentially choose whether to participate, knowing outcome of step 1 and decision of previous agents. *3* Pooled risk is split between s participating agents, each taking on risk P/t. Agents who do not participate in the reinsurance market take on risk $(1 - q_m)p_m$

2.3.2.3 Strategies

For elements $p \in \mathcal{P}$, we define the first-stage actions of all agents except agent m:

$$p_{-m} = (p_1, \ldots, p_{m-1}, p_{m+1}, \ldots, p_M) \in \prod_{m' \neq m} \mathcal{P}_{m'} \overset{\text{def}}{=} \mathcal{P}_{-m}.$$

A strategy for agent m consists of a pair: $A = (p_m, \eta_m) \in \mathcal{P}_m \times \{0, 1\}^{\mathcal{P}_{-m}}$, where p_m is the chosen portfolio of insurance, and $\eta_m : \mathcal{P}_{-m} \to \{0, 1\}$ is the participation choice, depending on the realization in the first stage.[21]

2.3.2.4 Belief Sets

Agent m has a belief set about the other agents' first-stage actions, $B_m = p_{-m} \in \mathcal{P}_{-m}$. Agent m's strategy, $A_m = (p_m, q_m)$, conditioned on belief set $B_m = p_{-m}$, is said to be *consistent*, if $\eta_m(p_{-m}) = (\mathcal{E}(\tilde{p}))_m$, where

$$\tilde{p} = ((p_{-m})_1, \ldots, (p_{-m})_{m-1}, p_m, (p_{-m})_{m+1}, \ldots, (p_{-m})_M), \qquad (2.9)$$

and we use the notation $(x)_i$ for the ith element of the ordered set x.

Rational agents will consider only consistent strategies, as inconsistent strategies are suboptimal in the participation phase of the diversification game. The inferred outcome of a consistent strategy, $A_m = (p_m, \eta_m)$, conditioned on a belief set, B_m, is

$$z_m(p_m | B_m) = \begin{cases} p_m, & \text{if } \eta_m(p_{-m}) = 0, \\ P/t, & \text{if } \eta_m(p_{-m}) = 1. \end{cases}$$

where

$$\tilde{q} = \mathcal{E}(\tilde{p}), \qquad t = \sum_{m'} (\tilde{q})_{m'}, \qquad P = \sum_{m'} (\tilde{p})_{m'} (\tilde{q})_{m'},$$

and $\tilde{p}$ is defined as in (2.9).

2.3.2.5 Equilibrium

An M-tuple of strategies, $(A_1, \ldots, A_M)$ and belief sets $(B_1, \ldots, B_M)$, where $A_m = (p_m, \eta_m)$ and $B_m = p_{-m}$, defines an equilibrium of the diversification game $\mathcal{G}$, if

---

[21] Here, in line with the previous discussion on reduced strategy space, $\tilde{q}_m$ does not need to be conditioned on the participation choices $q_{m'}$ of agents $m' = 1, \ldots, m-1$. This is the case as the equilibrium mapping $q = \mathcal{E}(p) \in \{0, 1\}^M$ is known, so $q_{m'}$ is uniquely implied by p in equilibrium.

1. Consistent strategies: For each agent, m, A_m is consistent, conditioned on belief set B_m.
2. Maximized strategies: For each agent, m, $p_m \in \arg\max_{p' \in P} U_m(z_m(p'|B_m))$.
3. Consistent beliefs: For each agent, m, for all $m' \neq m : (p_{-m})_{m'} = p_{m'}$.

The equilibrium outcome is summarized by $p = (p_1, p_2, \ldots, p_M)$ and $q = (\eta_1(p_{-1}), \eta_2(p_{-2}), \ldots,$
$\eta_M(p_{-M}))$. This concludes the definition of the diversification game.

The diversification game, of course, presents a highly stylized view of how primary markets and reinsurance markets for catastrophic risks work. A natural extension would be to allow the insurance premium (λ) to be defined endogenously by demand and supply. This extension turns out to complicate the analysis severely, so we have avoided it for analytical tractability. However, the nondiversification traps we derive occur for ranges of (fixed) λ's, so an interpretation of our result is that there may be no insurance premium, λ, for which there is both demand from potential insurance buyers and supply from single insurance providers.[22]

Another potential extension of the model would be to allow insurance providers to be able to take on fractions of risks, $x \in \mathcal{X}$ and not just 0 or 1. This type of extension would not qualitatively change our results, except for making the model less tractable.

2.3.3   Classification of Equilibria

We are interested in diversification and nondiversification equilibria to a diversification game $\mathcal{G} = (u, F, k, N, M)$. These formalize the situations that were intuitively described in Sect. 2.2. We define

Definition 2.3.1 A *diversification equilibrium* of a diversification game $\mathcal{G}$ is an equilibrium in which insurance against all risks in $\mathcal{X}$ is offered—i.e., $\overline{N} = NM$.

Definition 2.3.2 A diversification equilibrium of a diversification game $\mathcal{G}$ is *risk sharing* if all risk insured is pooled in the reinsurance market—i.e., $R = NM$.

Definition 2.3.3 A *nondiversification equilibrium* of a diversification game $\mathcal{G}$ is an equilibrium, in which no insurance against risk is offered—i.e., $\overline{N} = 0$.

---

[22]For example, in a calibration to earthquake insurance (see Ibragimov et al. 2009), we arrive at nondiversification trap arising for annual insurance premiums, λ, between USD 1,840 and USD 2,300 per household. Below USD 1,840, the only equilibrium is the nondiversification equilibrium, and above USD 2,300, the only equilibrium is the full-diversification equilibrium. The range of λ for which a nondiversification trap arises is thus about 20 % of the premium—i.e., (2,300–1,840)/2,340. With other parameter values, we have derived ranges from a few percent up to an order of magnitude.

Definition 2.3.4 A *nondiversification trap* exists in a diversification game $\mathcal{G}$, if there is both a nondiversification equilibrium and a risk-sharing diversification equilibrium.

We are especially concerned about cases when nondiversification traps may arise, even though there is a large risk-bearing capacity of the market as a whole. This might arise if the market is fragmented so coordination problems may be present— i.e., if M is large. We therefore define

Definition 2.3.5 A *genuine nondiversification trap* to the quadruple (u, F, k, N) exists if there exists an M_0, such that for all $M \geq M_0$, the diversification game $\mathcal{G} = (u, F, k, N, M)$ has a nondiversification trap.

In the next section, we analyze when traps can occur in the diversification game. It turns out that we can rigorously classify the conditions under which traps may occur.

2.3.4 Existence of Traps

We relate the equilibrium concepts described in Sect. 2.3.3 to conditions for the $U_{j,s}$ as defined in Eq. (2.7).

Condition 2.3.1 $U_{j,1} < U_{0,1}$ *for all* $j \in \{1, \ldots, N\}$.

Clearly, under Condition 2.3.1, an agent would never offer insurance if the reinsurance market were not available:

Condition 2.3.2 $U_{j,s} < U_{0,1}$ *for all* $j \in \{1, \ldots, N\}$ *and all* $s \in \{1, \ldots, M\}$.

Condition 2.3.2 is the stronger requirement that even if there is a reinsurance market, there is no way to increase expected utility by risk sharing if only one agent contributes risk to the reinsurance market. We shall see that a sufficient condition for there to be an equilibrium in which full diversification and risk sharing is achieved is

Condition 2.3.3

- $U_{NM,M} > U_{j,1}$ *for all* $j \in \{0, \ldots, N\}$ *and*
- $U_{NM,M} > U_{j,M}$ *for all* $j \in \{N(M-1), \ldots, NM-1\}$.

Our first set of results relates the existence of nondiversification traps to the expected utilities $\{U_{j,s}\}_{0 \leq j \leq NM, 1 \leq s \leq M}$, defined in (2.7). The results are fully in line with the arguments in Sect. 2.2. We have:

Proposition 2.3.1 *If Condition 2.3.2 is satisfied, then there is a nondiversification equilibrium.*

The implication can be almost reversed, as shown in

Proposition 2.3.2 *If Condition 2.3.2 fails strictly—i.e., if $U_{j,s} > U_{0,1}$ for some $j \in \{1, \ldots, N\}$ and $s \in \{1, \ldots, M\}$, then there is no nondiversification equilibrium.*

Proposition 2.3.3 *If Condition 2.3.3 is satisfied, then there is a risk-sharing diversification equilibrium.*

Clearly, if $U_{0,1} > U_{j,s}$ for all (j, s) such that $j \in \{1, \ldots, Ns\}$ and $s \in \{1, \ldots, M\}$, then the nondiversification equilibrium is unique. Under these conditions, the risks are by all means uninsurable, which may correspond to the "globally uninsurable" risks mentioned in Cummins (2006). Under such conditions, we can have no hopes for an insurance market to work: The risks are simply too large. Our analysis applies to situations for which risks may be "globally insurable," in that Condition 2.3.3 is satisfied but—in the terminology of Cummins (2006)— may be "locally uninsurable." In our model, local uninsurability is similar to Condition 2.3.1 being satisfied. For heavy-tailed distributions, Condition 2.3.2, which is stronger than Condition 2.3.1 may also be satisfied, which makes the "local uninsurability" especially cumbersome, and which may lead to coordination problems and nondiversification traps.

  We are now in a position to classify the situations when nondiversification traps can arise. We first show that genuine nondiversification traps can indeed be constructed. We have

Proposition 2.3.4 *In the model in Sect. 2.2, with Bernoulli–Cauchy distributions with parameters $N = 20$, $M = 5$, $\tilde{X} \in \tilde{S}^q_{\lambda,\mu,\sigma}$, with $\lambda = 1$, $\mu = -9$, $\sigma = 1$ and $q = 0.05$, $k = 100$, $u(x) = (x + k)^\alpha$, with $\alpha = 0.0315$, there is a nondiversification trap. Moreover, the nondiversification trap is genuine.*

As we shall see, the crucial point here is that the trap is *genuine* (the parameters in the example were not chosen to be empirically relevant). We next move on to classifying general distributional properties of the primitive risks that permit traps. It turns out that traps will arise only under quite specific conditions: First, nondiversification traps will not arise in a mean-variance framework with unlimited liability. Thus, in the traditional situation we will never see nondiversification traps. Second, genuine nondiversification traps can arise only with distributions that have heavy tails (i.e., infinite second moments).

Proposition 2.3.5 *If utility is of the form $Eu(X) = E(X) - \gamma Var(X)$, and $k = \infty$, then a nondiversification trap cannot occur. Moreover, depending on parameter values, only two situations can arise: Either there is a unique nondiversification equilibrium ($\overline{N} = 0$, $j = 0$, $t = 0$) or there is a unique diversification equilibrium with full risk sharing ($\overline{N} = NM$, $R = NM$, $t = M$).*

Non-genuine nondiversification traps can arise under standard conditions— i.e., distributions do not need to be heavy-tailed for nondiversification traps to be possible. For example, the diversification game $\mathcal{G} = (u, F, k, N, M)$, with

$$u(x) = xI(x \leq 0) + \log(1 + x)I(x > 0),$$
$$F(x) = I(x \geq -50)/2 + I(x \geq 70)/2,$$
$$k = \infty,$$
$$N = 20,$$
$$M = 5,$$

(where $I(\cdot)$ denotes the indicator function, and $F(x)$ thus is the cdf of a discrete r.v. X, with $P(X = -50) = P(X = 70) = 1/2$) has a nondiversification trap. However, *genuine* nondiversification traps arise only if distributions have heavy tails, as shown by the following proposition:

Proposition 2.3.6

 i) *If $k = \infty$ and the risks $X \in \mathcal{X}$ have finite second moments, i.e., $E(X^2) < \infty$, then a genuine nondiversification trap cannot occur.*
 ii) *If $k < \infty$, the risks $X \in \mathcal{X}$ have $E(X) \neq 0$ and $E(X^2) < \infty$ then a genuine nondiversification trap cannot occur.*
 iii) *If $k < \infty$, the risks $X \in \mathcal{X}$ have $E(X) = 0$ and $E(X^{2+\epsilon}) < \infty$, for some arbitrary small $\epsilon > 0$, then a genuine nondiversification trap cannot occur.*

Proposition 2.3.6 can also be viewed from an approximation perspective. If M is large, but finite, then nondiversification traps can arise only with distributions that have left tails that are "approximately" heavy—i.e., decay slowly up until a certain point (even though their real support may be bounded). For details on this type of argument, see Ibragimov and Walden (2007).

2.3.5 Traps in Markets for Catastrophic Insurance

We apply our results to real markets for catastrophic insurance. Obviously, risks vary across product lines and geography, and a full investigation is outside the scope of this monograph. Instead, we focus on one type of risk—earthquake insurance in California. Applying the principles of seismology, we show that the distribution of loss sizes indeed follows a heavy tailed power law and that an exponent of unity (the Cauchy case) is by no means unreasonable. Moreover, with a simple calibration, we estimate the value of being able to avoid a trap in residential earthquake insurance in California to be up to USD 3.0 billion per year. This is the direct value effect of a trap. The estimate does not include indirect effects, as, e.g., analyzed in Hubbard et al. (2005). We also discuss how this type of analysis is valid for other types of natural disasters. Finally, we relate our results to several recent events in markets for catastrophic insurance.

2.3.5.1 Loss Distribution of Natural Disasters

The fact that earthquakes are referred to as catastrophes is suggestive that they have heavy-tailed distributions. In this section, we show more precisely that standard seismic theory leads to loss distributions that follow Pareto laws $h_L(l) \sim l^{-\xi}$ in (1.1). Here, $h_L(l) = P(L > l)$ is the probability that the economic loss, L, is larger than l, conditioned on an earthquake occurring. More generally, for an r.v. X, let $h_X(x)$ denote the probability that X exceeds x, conditioned on an earthquake occurring, $P(X \geq x)$.

Pareto laws arise for the distributions of energy release from earthquakes (see, e.g., Sornette et al. 1996). We show that *economic loss* also satisfies a Pareto law. For economic loss estimates, it is more natural to work with the Modified Mercalli Intensity (MMI) scale. Let M denote the *moment magnitude* Hanks and Kanamori (1979) of an earthquake.[23] A standard model for the distribution of moment magnitudes of earthquakes is

$$h_M(m) = C_1 e^{-\beta m}, \tag{2.10}$$

where $\beta = 1.84$ is often used (see McGuire 2004, pp. 34–40).[24] The exponential distribution is adequate for $M \leq 7$, but for higher M, it *underestimates* the probabilities (McGuire 2004, pp. 53–54; Schwartz and Coppersmith 1984), so the distribution for high levels may in fact have heavier tails than assumed in (2.10).[25]

An empirical relationship between the MMI and the expected magnitude is given by

$$M = 1.3 + 0.6 I_e \quad \Rightarrow \quad I_e = \frac{M}{0.6} - \frac{1.3}{0.6}, \tag{2.11}$$

where I_e is the epicentral intensity—i.e., the MMI at the center of the earthquake (McGuire 2004, p. 44). For simplicity, we assume that this is a deterministic relationship. The MMI at a specific point is directly related to the damage and losses at that point. For example, for an MMI of 8, the estimates of losses for wooden

---

[23]The moment magnitude is almost the same as the Richter magnitude, M_R, for $M \leq 6.5$, but provides a more accurate measure for earthquakes of larger magnitudes.

[24]This is the moment magnitude version of the celebrated Gutenberg–Richter exponential law for the Richter magnitude.

[25]Although for *very* high levels, physical arguments imply that there has to be an upper bound on the energy released; see Knopoff and Kagan (1977), and Kagan and Knopoff (1984). However, even if there is an upper bound, say at $M = 10$ to 11, this still leads to an approximate Pareto law for over 15 magnitudes of energy release. The upper bound is well beyond the limited liability threshold of most insurance markets and is therefore not crucial for our trap argument.

structures is 5–10 % of total value (McGuire 2004, p. 19).[26] For I_e, we immediately get[27]

$$h_{I_e}(i) = C_1 e^{-\beta(1.3+0.6i)} = C_2 e^{-1.10i}.$$

We relate the area, A, covered by an earthquake to I_e through the *attenuation function*. We use the estimate

$$I_d = I_e + 2.87 + 0.00052D - 1.25 \log_{10}(D+10) \geq I_e + 2.87 - 1.25 \log_{10}(D+10),$$

where I_d is the MMI at a point of distance D away from the epicenter; see Ho et al. (2001).[28] Let $A_d(I_e, I_d)$ denote the area that experiences an MMI $\geq I_d$ for an earthquake with epicentral intensity I_e. We also write $A(I_e)$ when I_d is fixed and known. Then, as $A \sim D^2$, it is easy to see that

$$A(I_e, I_d) \geq C_3 \times 10^{1.6(I_e-I_d)} = C_3 \times e^{1.6 \ln(10)(I_e-I_d)} = C_4 \times e^{3.7I_e},$$

for fixed I_d. Another estimate is obtained by using the results in Hanks and Johnston (1992), for $I_d = 6$. Their formula is

$$M = 2.38 + 0.96 \log_{10}(A(I_e(M), 6))$$

which by (2.11) leads to

$$A(I_e, 6) = C_5 \times e^{0.6 \times \ln(10)I_e/0.96} = C_5 e^{1.44I_e}.$$

Under the assumption of uniform geographical population density, it is natural to assume that the economic loss, L, from an earthquake is *at least* proportional to $A(I_e)$, as this estimate only takes into account area covered, but does not take into account that the higher I_e, the more damages occur close to the epicenter. This leads to the loss distribution[29]:

$$h_L(l) \sim l^{-\zeta}, \qquad \text{where } \zeta \in [0.3, 0.76].$$

---

[26]This estimate may be somewhat outdated, as building structures nowadays may be stronger. However, this does not change our general conclusions, only the constants in the formulae (personal communication with William L. Ellsworth, Chief Scientist, Western Region Earthquake Hazards Team, United States Geological Survey).

[27]$1.10 \approx 1.84 \times 0.6$.

[28]Other estimates for the relation are available—e.g., in Bakun et al. (2003). However, as with the strength of building structures, they are qualitatively similar and will not change our main conclusions (personal communication with William L. Ellsworth, Chief Scientist, Western Region Earthquake Hazards Team, United States Geological Survey).

[29]$0.3 \approx 1.10/3.7$, $0.76 \approx 1.10/1.44$.

The other extreme assumption is that of one-dimensional population density—i.e., that people only live along a one-dimensional coastline. In this case it is natural to assume that the economic loss, L, is *at least* proportional to $\sqrt{A(I_e)}$. This leads to the loss distribution[30]:

$$h_L(l) \sim l^{-\zeta}, \qquad \text{where } \zeta \in [0.6, 1.5].$$

Thus, altogether the economic loss distribution follows a power law with tails that decay slower than $\zeta = 1.5$. The corresponding tail index may be as slow as $\zeta = 0.3$.

Clearly, these calculations are rough. However, the key point is that under standard assumptions, for many orders of magnitude, the distribution of economic loss from earthquakes follows an approximate power law, with a very heavy tail and that the tail exponent $\zeta = 1$ by no means is unreasonable.

The crucial property for our theory is the heavy-tailed distributions, especially power laws like in (1.1). These are generic for natural disasters. As discussed in Woo (1999), heavy-tailedness is intimately connected to the self-similarity of the physical processes underlying natural disasters. For example, the energy distribution released in earthquakes satisfies a power law with an exponent between $\zeta \in (0.8, 1.2)$ (Sornette et al. 1996). Similarly, the energy distribution of extraterrestrial impacts (meteorites and asteroids) satisfies a power law with exponent $\zeta \approx 0.86$, the size distribution of landslides has been estimated to have an exponent of $\zeta \in (1.2, 1.4)$, whereas the area covered by river floods scales with the exponent $\zeta \approx 0.43$ (Woo 1999). For hurricanes in Florida, estimates of the tail of the loss distribution of $\zeta \approx 1.56$ (Hsieh 1999) and $\zeta \approx 2.49$ (Hogg and Klugman 1983) have been made.

In each of these cases, a relationship between the heavy-tailed variable, X, and economic loss, L, needs to be established, just like when going from moment magnitude to economic loss for earthquakes. However, as long as this relationship is of power-type, $L(X) \sim X^\beta$, the loss distribution will also satisfy a power law, with the tail index $\zeta = \zeta/\beta$, so our theory can be applied.

2.3.5.2   The Role of a Central Agency

As the private catastrophe insurance markets for earthquakes, wind damage, floods, and terrorism have failed, one after the other over the past 15 years, in both the USA and Europe, governments have been forced to intervene. The plans were often created under time pressure and they differ substantially in their details; see OECD (2005a,b) for descriptions of both the European and U.S. plans. So it is intriguing to find that they actually share a fundamental design feature—namely, that each government plan has, in effect, created a mechanism through which a coordinated diversification equilibrium is established.

---

[30]$0.6 \approx 1.10/(3.7/2)$, $1.5 \approx 1.10/(1.44/2)$.

For example, in the USA following the terrorism attack of September 11, 2001, the U.S. Congress passed the TRIA, which requires all U.S. insurance firms to offer terrorism coverage as a rider to their standard coverage for commercial buildings. The quid pro quo is that the government provides reinsurance for the highest layer of risk, although, as shown in Carroll et al. (2005), the actual subsidy is very small. Thus, the primary force of TRIA is that it requires a coordinated equilibrium in which all insurers must offer terrorism coverage. The federal government also directly provides most U.S. flood insurance, which, of course, automatically diversifies the risk across all U.S. taxpayers. At the state level, Florida and California have required private firms to continue to cover hurricane and earthquake risks respectively, while the states support some of the reinsurance. Finally, most European countries have created national catastrophe programs covering both natural disasters and terrorism that generally require that catastrophe coverage be offered to all customers, while the government provide a reinsurance facility; see OECD (2005a,b).

Thus, quite systematically, government interventions to support catastrophe insurance markets in both the USA and Europe have, in effect, created coordinated diversification equilibria. This supports the main conclusion of this section, that government support to help reach a coordinated diversification equilibrium may play an important role in maintaining functioning markets for catastrophe insurance.

2.4 Financial Intermediation: Diversification Disasters[31]

Which factors determine the risks of systemic failures of financial institutions and the benefits of diversification? When do the risks outweigh the benefits? What are the policy implications of such a trade-off? In this section, we analyze these questions in a parsimonious model. In the model, while individual institutions may have an incentive to diversify their risks, diversification creates a negative externality in the form of systemic risk. If all intermediaries are essentially holding the same diversified portfolio, a shock may disrupt all the institutions simultaneously, which is costly to society, since it may take time for the financial system, and thereby the economy, to recover. Specifically, the slow recovery time creates a significant and continuing social cost because the unique market-making and information analysis provided by banks and other intermediaries[32] is lost until they recover; see Bernanke (1983). Indeed, Bernanke's concern with the social cost created by bank failures appears to have motivated many of the government bank bailouts.

---

[31]This section is based on the article Ibragimov et al. (2011), which was published in the *Journal of Financial Economics*, Vol. 99, Issue 2, pp. 333–348, Copyright Elsevier (2010).

[32]Our analysis applies to banks, but more broadly to general financial intermediaries, like pension funds, insurance companies, and hedge funds.

In our model, the costs and benefits of risk-sharing are functions of five properties of the economy. First, the number of asset classes is crucial: The fewer the number of distinct asset classes that are present, the weaker the case for risk-sharing. Second and third, the dependence between risks within an asset class, and the heavy-tailedness of the risks are important. The larger the positive dependence (i.e., higher correlation, if defined) and the heavier the tails of the risk distribution, the less beneficial risk-sharing is. Fourth, the longer it takes for the economy to recover after a systemic failure, the more costly risk-sharing is and, fifth, lower discount rates also work against risk-sharing. We define the *diversification threshold* to be the threshold at which the cost to society of systemic failure begins to exceed the private benefits of diversification, and we derive a formula for the threshold as a function of these five properties.

The distributions of the risks that intermediaries take on are key to our results. When these risks are thin-tailed, risk-sharing is always optimal for both individual intermediaries and society. But, with moderately heavy-tailed risks, risk-sharing may be suboptimal for society, although individual intermediaries still benefit from it. In this case, the interests of society and intermediaries are unaligned. For extremely heavy-tailed risks, intermediaries and society once again agree, this time that risk-sharing is suboptimal.

Our analysis has implications for risk management and policies to mitigate systemic externalities. We show that VaR considerations lead individual intermediaries to diversify, as per incentives similar to those in the Basel bank capital requirements. Within our framework, however, the diversification actions may lead to suboptimal behavior from a societal viewpoint. It then becomes natural to look for devices that would allow individual firms to obtain the benefits of diversification, but without creating a systemic risk that could topple the entire financial system. In Sect. 2.4.2, we provide a framework to develop such solutions and provide specific proposals.

This section of the book is related to the recent, rapidly expanding literature on systemic risk and market crashes. The closest paper is Acharya (2009). Our definition of systemic risk is similar to Acharya's, as are the negative externalities of joint failures of intermediaries. The first and foremost difference between the two works is our focus on the distributional properties of risks and the number of risk classes in the economy, which is not part of the analysis in Acharya (2009). Moreover, the mechanisms that generate the systemic risks are different in the two works. Whereas the systemic risk in Acharya (2009) arises when individual intermediaries choose correlated real investments, in our model the systemic risk is introduced when intermediaries with limited liability become interdependent when they hedge their idiosyncratic risks by taking positions in what is in effect each others' risk portfolios. Such interdependence may have been especially important for systemic risk in the recent financial crisis. This leads to a distinctive set of policy implications, as we develop in Sect. 2.4.2.

Wagner (2010) independently develops a model of financial institutions in which there are negative externalities of systemic failures, and diversification therefore may be suboptimal from society's perspective. Wagner's analysis, however, focuses on the effects of conglomerate institutions created through mergers and acquisitions and the effects of contagion. Furthermore, the intermediary size and investment decisions are exogenously given in Wagner's study, and only a uniform distribution of asset returns is considered. Our model, in contrast, emphasizes the importance of alternative risk distributions and the number of risks in determining the possibly negative externality of diversification. The two studies therefore complement each other.

A related literature models market crashes based on contagion between individual institutions or markets. A concise survey is available in Brunnermeier (2009). Various propagation mechanisms have been used, typically through an externality in which the failure of some institutions triggers the failure of others. Rochet and Tirole (1996) model an interbank lending market, which intrinsically propagates a shock in one bank across the banking system. Allen and Gale (2000) extend the Diamond and Dybvig (1983) bank run liquidity risk model, such that geographic or industry connections between individual banks, together with incomplete markets, allow for shocks to some banks to generate industry-wide collapse. Kyle and Xiong (2001) focus on cumulative price declines that are propagated by wealth effects from losses on trader portfolios. Kodes and Pritsker (2002) use informational shocks to trigger a sequence of synchronized portfolio rebalancing actions, which can depress market prices in a cumulative fashion. Caballero and Krishnamurthy (2008) focus on Knightian uncertainty and ambiguity aversion as the common factor that triggers a flight to safety and a market crash. Most recently, Brunnermeier and Pedersen (2009) model a cumulative collapse created by margin requirements and a string of margin calls.

The key commonalities between the present section of the book and this literature is the possibility of an outcome that allows a systemic market crash, with many firms failing at the same time. Moreover, as in many other papers, in our model there is an externality of the default of an intermediary—in our case, the extra time it takes to recover when many defaults occur at the same time. The key distinction between this section and the above literature is, again, our focus on the importance of risk distributions and number of asset classes in an economy. Thus, a unique feature of our model is that the divergence between private and social welfare arises from the statistical features of the loss distributions for the underlying loans alone. This leads to strong, testable implications and to distinctive policy implications. In our model, these effects arise even without additional assumptions about agency problems (e.g., asymmetric information) or third-party subsidies (e.g., government bailouts). No doubt, such frictions and distortions would make the incentives of intermediaries and society even less aligned.

2.4.1 Model

We use the following conventions: lowercase thin letters represent scalars, uppercase thin letters represent sets and functions, lowercase bold letters represent vectors, and uppercase bold letters represent matrices. The ith element of the vector

$\mathbf{v}$ is denoted $(\mathbf{v})_i$, or $\mathbf{v}_i$ if this does not lead to confusion, and the n scalars v_i, $i = 1, \ldots, n$ form the vector $[v_i]_i$. We use T to denote the transpose of vectors and matrices. One specific vector is $\mathbf{1}_n = \underbrace{(1, 1, \ldots, 1)}_{n}^{T}$, (or just $\mathbf{1}$ when n is obvious).

Similarly, we define $\mathbf{0}_n = \underbrace{(0, 0, \ldots, 0)}_{n}^{T}$.

Consider an infinite horizon economy, $t \in \{0, 1, 2, \ldots\}$, in which there are M different risk classes. Time value of money is represented by a discount factor $\delta < 1$ so that the present value of one dollar at $t = 1$ is δ. There is a bond market in perfectly elastic supply, so that at t, a risk-free bond that pays off one dollar at $t + 1$ costs δ.

There are M risk-neutral trading units, each trading in a separate risk class. We may think of unit m as a representative trading unit for risk class m. Henceforth, we shall call these trading units *intermediaries*, capturing a large number of financial institutions, like banks, pension funds, insurance companies, and hedge funds. We thus assume that each trading unit, or intermediary, specializes in one risk class. Of course, in reality, intermediaries hold a variety, perhaps a wide variety of risks. The key point here is that the intermediaries are not initially holding the market portfolio of risks, so that they may have an incentive to share risks with each other.

We think of the M risk classes as different risk lines or "industries," e.g., representing real estate, publicly traded stocks, private equity, etc. Within each risk class, in each time period t, there is a large number, N, of individual multivariate normally distributed risks, $x_n^{t,m}$, $1 \leq n \leq N$. We will subsequently let N tend to infinity, whereas M will be a small constant, typically less than 50, as is typical in financial and insurance applications (the results in this section also hold in the case when the number of risks, N, in the mth class depends on m, as long as we let the number of risks in each risk class tend to infinity). For simplicity, we assume that risks belonging to different risk classes are independent, across time t, and risk class, n i.e., $x_n^{t,m}$ is independent of $x_{n'}^{t',m'}$ if $m \neq m'$ or $t' \neq t$. This is not a crucial assumption; similar results would arise with correlated risk classes. For the time being, we focus on the first risk class, in time period zero. We therefore drop the m and t superscripts.

We make some stylized assumptions about how risks within a risk class are related. The idea is to introduce a "distance" between risks, so that some risks are closer—and thereby more related—than others. For simplicity, we make strong assumptions about the form of these risk dependencies, namely that they have a one-dimensional circulant topological structure.[33] These assumptions will allow us to provide a qualitative characterization of the interactions between uncertainty, heavy-tails, and limits to diversification.

---

[33]Circulant topological structures have been used in the economics literature to provide a simple spatial "distance" metric without discontinuities, see, e.g., the discussion in Hennessy and Lapan (2009) and the references therein.

Specifically, per assumption, the individual risks have multivariate normal distributions, related by

$$x_{i+1} = \rho x_i + w_{i+1}, \qquad i = 1, \dots, N-1, \tag{2.12}$$

for some $\rho \in [0, 1)$.[34]

Here, w_i are independent and identically distributed (i.i.d.) normally distributed random variables with zero mean and variance $\sigma^2(w_i) = 1 - \rho^2$. Each x_i represents cross-sectional risk, with local dependence in the sense that $\text{cov}(x_i, x_j)$ quickly approaches zero when $|i - j|$ grows, i.e., the decay is exponential. The risks could, for example, represent individual mortgages and the total risk class would then represent all the mortgages in the economy. For low ρ, the risks of these mortgages are effectively uncorrelated, except for risks that are very close. "Close" here could, for example, represent mortgages on houses in the same geographical area. If ρ is close to one, shocks are correlated across large distances, e.g., representing country-wide shocks to real estate prices. This structure thus allows for both "local" and "global" risk dependencies in a simple setting.[35]

For simplicity, we introduce symmetry in the risk structure by requiring that

$$x_1 = \rho x_N + w_1, \tag{2.13}$$

i.e., the relationship between x_1 and x_N is the same as that between x_{i+1} and x_i, $i = 1, \dots, N-1$. This choice of risk structure conveniently implies that all x_is have standard normal marginal distributions. We can rewrite the risk structure in matrix notation, by defining $\mathbf{x} = [x_i]_i$, $\mathbf{w} = [w_i]_i$. The relationship (2.12,2.13) then becomes

$$\mathbf{Ax} = \mathbf{w}.$$

Here, $\mathbf{A}$ is an invertible so-called circulant Toeplitz matrix,[36] given by

$$\mathbf{A} = \text{Toeplitz}_N[-\rho, \underline{1}, \mathbf{0}_{N-2}^T, -\rho].$$

---

[34]We focus on multivariate normal risks, for tractability. Similar results arise with other, thin-tailed, individual risks, e.g., Bernoulli distributions, although the analysis becomes more complex, because other distribution classes are not closed under portfolio formation so the central limit theorem needs to be incorporated into the analysis.

[35]For review and discussion of models with common shocks and modeling approaches for spatially dependent economic and financial data, see, among others, Conley (1999), Andrews (1993), Ibragimov and Walden (2007), and Ibragimov (2009b).

[36]A Toeplitz matrix $\mathbf{A} = \text{Toeplitz}_N[a_{-N+1}, a_{-N+2}, \dots, a_{-1}, a_0, a_1, \dots, a_{N-2}, a_{N-1}]$, is an $N \times N$ matrix with the elements given by $(\mathbf{A})_{ij} = a_{j-i}$, $1 \leq i \leq N$, $1 \leq j \leq N$. A Toeplitz matrix is banded if $(A)_{ij} = 0$ for large $|j - i|$, corresponding to $a_i = 0$ for indices i that are large by absolute value. When $a_i = 0$ if $i < -k$ or $i > m$, for $k < N - 1$ or $m < N - 1$, we use the notation $a_{-k}, a_{-k+1}, \dots, a_{\underline{0}}, \dots, a_{m-1}, a_m$ to represent the whole sequence generating the Toeplitz matrix. For example, the notation $\mathbf{A} = \text{Toeplitz}_N[a_{-1}, a_{\underline{0}}]$ then means that $\mathbf{A}_{ii} = a_0$, $\mathbf{A}_{i,i-1} = a_{-1}$, and that all other elements of $\mathbf{A}$ are zero. For an $N \times N$ Toeplitz matrix, if $a_{N-j} = a_{-j}$, then the matrix is, in addition, circulant. See Horn and Johnson (1990) for more on the definition and properties of Toeplitz and circulant matrices.

Thus, given the vector with independent noise terms, $\mathbf{w}$, the risk structure, $\mathbf{x}$, is defined by

$$\mathbf{x} \stackrel{\text{def}}{=} \mathbf{A}^{-1}\mathbf{w}.$$

The symmetry is merely for tractability and we would expect to get similar results without it (although at the expense of higher model complexity). In fact, for large N, the covariance structure in our model is very similar to that of a standard AR(1) process, defined by

$$\hat{x}_0 = \hat{w}_0,$$
$$\hat{x}_{i+1} = \rho\hat{x}_i + \hat{w}_{i+1}, \qquad i = 0, \ldots, N-1,$$

where the $\hat{w}_i$'s are independent, $\sigma^2(\hat{w}_0) = 1$, $\sigma^2(\hat{w}_i) = 1 - \rho^2$, $i > 0$ (although, of course, i does not denote a time subscript in our model, as it does in a standard AR(1) process). In this case, the matrix notation becomes $\hat{\mathbf{A}}\hat{\mathbf{x}} = \hat{\mathbf{w}}$, where $\hat{\mathbf{A}} = \text{Toeplitz}_N[-\rho, 1]$. The only difference between $\mathbf{A}$ and $\hat{\mathbf{A}}$ is that $\mathbf{A}_{1N} = -\rho$, whereas $\hat{\mathbf{A}}_{1N} = 0$. The covariance matrix $\hat{\Sigma} = [\text{cov}(\hat{x}_i, \hat{x}_j)]_{ij}$ has elements

$$\hat{\Sigma}_{ij} = \rho^{|i-j|}.$$

Thus, in the AR(1) setting, the covariance between risks decreases geometrically with the distance $|i - j|$. As we shall see in Theorem 2.4.1, this property carries over to the covariances in our symmetric structure when the number of risks within an asset class, N, is large, (although, technically, for large N the distance needed to be interpreted as the distance on the circle, i.e., the distance between i and j is $\min(|j - i|, N - |i - j|)$).

We assume that the correlations between individual risks (2.12) are uncertain. Specifically,

$$P(\rho \leq y) = 1 - \left(\frac{1-y}{1+y}\right)^\gamma, \qquad y \in [0, 1), \ \gamma > 0, \tag{2.14}$$

i.e., the probability that the correlation is close to one satisfies a power-type law with the tail index γ, where a low γ indicates that there is a substantial chance that correlations are high, whereas a high γ indicates that correlations are very likely low. The uncertainty of correlations will lead to heavier-tailed distributions for *portfolios* of risks, just as uncertain variances may cause the distribution of an *individual* risk to have a heavy tail.[37]

---

[37] As we shall see, our approach leads to very tractable formulas. An alternative approach for introducing more complex correlation structures than the standard multivariate normal one is to use copulas, which may also lead to heavy-tailed portfolio distributions.

Relation (2.14) is equivalent to the condition that the ratio $\frac{1+\rho}{1-\rho}$ follows the Pareto distribution in (1.1) with the tail index $\zeta = \gamma$:

$$P\left(\frac{1+\rho}{1-\rho} \geq v\right) = v^{-\gamma}, \qquad v \geq 1. \tag{2.15}$$

We first consider the case in which intermediaries do not trade risks with each other. At $t = 0$, an intermediary invests in a portfolio of x_i unit risks, with the total portfolio size determined by the vector $\mathbf{c} \in \mathbf{R}^N$. The total portfolio is then $\mathbf{c}^T\mathbf{x}$.[38] Here, $\mathbf{c}_i$ represents the number of units of $\mathbf{x}_i$ risk that the intermediary invests in. The total dollar outcome of the investment in risk i, after the risks are realized, is $c_i x_i$ and the total dollar value of the whole portfolio, after realization, is $\mathbf{c}^T\mathbf{x}$. Since the elements of $\mathbf{x}$ can take on negative values, it is natural to think of $\mathbf{c}^T\mathbf{x}$ as containing both long and short positions. The short positions may be in risk-free capital (i.e., leverage through debt), but also in risky assets. At $t = 0$, the intermediary also reserves capital, so that $k \leq K$ is available at $t = 1$, where K represents the maximum capital (in dollar terms) available for the intermediary to reserve and is exogenously given. For simplicity, we assume that the capital is invested in risk-free cash. We could allow the capital to be invested in the risky portfolio, but this would further complicate the formulas without any new insights. The intermediary has limited liability, so its losses are bounded by k. If the intermediary defaults, the additional losses are borne by the counter-party. The shortfall that may be imposed on the counter-party is taken into account in the asset pricing.

At $t = 1$, the values of the $t = 0$ risks are realized. Because of limited liability, the value of the portfolio (in dollars) is then

$$\max(k + \mathbf{c}^T\mathbf{x}, 0) \stackrel{\text{def}}{=} \mathbf{c}^T\mathbf{x} + k + Q, \tag{2.16}$$

where $Q = \max(k + \mathbf{c}^T\mathbf{x}, 0) - \mathbf{c}^T\mathbf{x} - k$ is the realized value of the option to default (see Ibragimov et al., 2012). The intermediary ensures that the amount of capital available at $t = 1$ is k. The value of the portfolio of investments at $t = 1$, above the capital reserved, is therefore $\mathbf{c}^T\mathbf{x} + Q$.

The price the intermediary pays at $t = 0$ for this portfolio of investments is its discounted expected value, $\delta E[\mathbf{c}^T\mathbf{x} + Q]$, less a premium, d, per unit risk. This premium represents the part of the surplus generated by the transaction that is captured by the intermediary. A similar feature is used in the work of Froot et al. (1993) and Froot and Stein (1998). In a full equilibrium model, d would be endogenously derived, but for tractability we assume that it is exogenously given and that it is constant.[39]

---

[38]We allow for short-selling. In the proof, we show that the optimal portfolio does not involve short-selling, so we could equivalently have permitted only nonnegative portfolios, $\mathbf{c} \in \mathbf{R}^N_+$.

[39]In Ibragimov et al. (2010), equilibrium premiums are derived in a model with multiple risk factors and risk-averse agents. The general model, however, is quite intractable, and the simplifying

If the intermediary defaults, it is out of business from there on, generating zero cash flows in all future time periods.[40] Later, we will introduce the possibility for the intermediaries to trade with each other, after $t = 0$ (when they take on the risk), but before $t = 1$ (when the value of the risk is realized), but for the time being, we ignore this possibility.

The ex ante value of the total cash flows to the owners of the intermediary between $t = 0$ and $t = 1$ is then

$$\underbrace{d\left(\mathbf{c}^T\mathbf{1}\right) - \delta k - \delta E[\mathbf{c}^T\mathbf{x} + Q]}_{t=0} + \underbrace{\delta E[(k + \mathbf{c}^T\mathbf{x} + Q)]}_{t=1} = d\left(\mathbf{c}^T\mathbf{1}\right). \qquad (2.17)$$

This is thus the net present value of operating the intermediary between $t = 0$ and $t = 1$.

Now, if the intermediary survives, which it does with probability

$$q \stackrel{\text{def}}{=} P(-\mathbf{c}^T\mathbf{x} \le k),$$

the situation is repeated, i.e., at $t = 1$ the intermediary takes on new risk, reserves capital, and then at $t = 2$ risks are realized, etc. If the intermediary defaults at any point in time, it goes out of business and its cash flows are zero from there on. In the infinite horizon case, the value of the intermediary in recursive form is therefore

$$V^B = d\left(\mathbf{c}^T\mathbf{1}\right) + \delta q V^B,$$

implying that

$$V^B = \frac{d\left(\mathbf{c}^T\mathbf{1}\right)}{1 - \delta q}. \qquad (2.18)$$

We note that the counter-party of the $\mathbf{x}$ risk transaction understands that the intermediary may default, and takes this into account when the price for the risk is agreed upon. Therefore, since the price of the contract takes into account the risk level of the intermediary, the counter-party does not need to impose additional covenants.

---

assumptions in this section allow us to carry through a more complete study of the role of risk distributions.

[40]It may be optimal, if possible, for the owners of the intermediary to infuse more capital even if losses exceed k, to keep the option of generating future profits alive. Equivalently, they may be able to borrow against future profits. Taking such possibilities into account would increase the point of default to a value higher than k, but would qualitatively not change the results, since there would always be some realized loss level beyond which the intermediary would be shut down even with such possibilities.

Equation (2.18) describes the trade-off the individual intermediary makes when choosing its portfolio. On the one hand, a larger portfolio increases the cash flows per unit time—increasing the numerator—but on the other hand, it also increases the risk of default—increasing the denominator. It is straightforward to show that if the intermediary could, it would take on an infinitely large portfolio. This is shown in Ibragimov et al. (2011). In terms of Eq. (2.18), the numerator effect dominates the denominator effect.

A regulator, representing society, therefore imposes restrictions on the probability for default, to counterweight the risk-shifting motive. Specifically, the intermediary's one-period probability of default is not allowed to exceed β at any point in time. In other words, the intermediary faces the constraint that the VaR for the loss probability of β can be at most k, $\text{VaR}_\beta(\mathbf{c}^T\mathbf{x}) \leq k$. Therefore, since the realized dollar losses are $\mathbf{c}^T\mathbf{x}$ and the capital reserved is k, the VaR constraint says that the probability is at most β that the realized losses exceed the capital. In our notation, this is the same as to say that $q \geq 1 - \beta$.

The VaR constraint is imposed in the model to reflect the existing management and regulatory standards through which most financial intermediaries currently operate. Most major financial intermediaries apply VaR as a management tool and regularly report their VaR values. The Basel II bank capital requirements are based on VaR. European securities firms face the same Basel II VaR requirements, while the major U.S. securities firms have become bank holding companies and thus also face these requirements. European insurance firms are being reregulated under "Solvency II," which is VaR-based, in parallel to Basel II. U.S. insurers are regulated by individual states with capital requirements that vary by state and insurance line; these are generally consistent with a VaR interpretation.

To be clear, VaR is not necessarily the optimal risk measure when financial intermediary behavior may create systemic externalities. Indeed, in this section we show that VaR requirements can lead to diversification decisions by individual intermediaries that are inconsistent with maximizing the societal welfare. In Sect. 2.4.2, we consider alternative proposals for intermediary regulation when systemic externalities are important.

It is natural to think of $\frac{\mathbf{c}^T\mathbf{1}}{k}$ as a measure of the leverage of the firm, since $\mathbf{c}^T\mathbf{1}$ represents the total liability exposure and k represents the capital that can be used to cover losses.

The assumption that there is an upper bound, K, that the intermediary faces on how much capital can be reserved is reduced form. In a full equilibrium model without frictions, the investment level would be chosen such that the marginal benefit and cost of an extra dollar of investment would be equal, and K would then be endogenously derived. In our reduced-form model, where the benefits d are constant, we assume that the marginal cost of raising capital beyond K is very high, so that K provides an effective hard constraint on the capital raising abilities of the intermediary. We note that the bound may be strictly lower than the friction-free outcome, because of other frictions on capital availability in imperfect financial markets (see, e.g., Froot et al. 1993 for a discussion of such frictions). Given a choice of capital, $k \leq K$, the VaR constraint imposed by the regulator then imposes a bound on how aggressively the intermediary can invest, i.e., on the intermediary's size. The VaR constraint therefore also automatically imposes a capital requirement restriction on the intermediary. We next study the intermediary's behavior in this environment.

2.4.1.1 Optimal Behavior of Intermediary

In line with the previous arguments, the program for the intermediary is

$$\max_{c,k} \frac{d\,(\mathbf{c}^T \mathbf{1})}{1 - \delta P(-\mathbf{c}^T \mathbf{x} \leq k)}, \qquad \text{s.t.,} \tag{2.19}$$

$$k \in [0, K], \tag{2.20}$$

$$\mathbf{c} \in \mathbf{R}^n, \tag{2.21}$$

$$k \geq \mathrm{VaR}_\beta(\mathbf{c}^T \mathbf{x}). \tag{2.22}$$

The following theorem characterizes the intermediary's behavior:

Theorem 2.4.1 *Given a VaR constrained intermediary solving (2.19)–(2.22), where the risks are of the form (2.12), (2.13), β is close to zero, and the distribution of correlations is of the form (2.14). Then, for large N:*

a. For a given ρ, the covariance $cov(x_i, x_j)$ converges to $\rho^{|i-j|}$ for any i, j.
b. The payoff of the intermediary's chosen risk portfolio, $\mathbf{c}^T \mathbf{x}$, follows power law (1.2) in the left tail with the tail index $\xi = 2\gamma$, and with the pdf $f(y)$, where

$$f(y) = \frac{\gamma 2^\gamma b^{-\gamma}}{\sqrt{\pi}} \times \frac{\Gamma\left(\gamma + \frac{1}{2}, \frac{by^2}{2}\right)}{|y|^{1+2\gamma}}, \tag{2.23}$$

for some constant, $b > 0$, for all $y \in \mathbf{R} \backslash \{0\}$. Here, Γ is the lower incomplete Gamma function, $\Gamma(a, x) \stackrel{\text{def}}{=} \int_0^x t^{a-1} e^{-t} dt$, $a > 0$. Also, $f(0) = \frac{2\gamma}{b\sqrt{2\pi}(2\gamma+1)}$.
c. Maximal capital is reserved, $k = K$, i.e., (2.20) is binding.
d. Maximal VaR is chosen, i.e., (2.22) is binding.
e. If $\gamma > 1$, then the variance of the portfolio is

$$\sigma^2 = b^2 \frac{\gamma}{\gamma - 1}, \tag{2.24}$$

else it is infinite.

We note that, since $\Gamma(a, x) = \Gamma(a) + o(1)$, as $x \to +\infty$, (where $\Gamma(a)$ is the Gamma function, $\Gamma(a) = \int_0^\infty t^{a-1} e^{-t} dt$, $a > 0$), we obtain that the distribution of $\mathbf{c}^T \mathbf{x}$ has a heavy power law left tail (1.2):

$$P(-\mathbf{c}^T \mathbf{x} > y) = \frac{2^\gamma b^{-\gamma} \Gamma\left(\gamma + \frac{1}{2}\right) + o(1)}{\gamma \sqrt{\pi} |y|^{2\gamma}}, \quad y \to \infty.$$

Thus, the portfolio chosen by the intermediary when solving (2.19)–(2.22) is moderately heavy-tailed, or even heavy-tailed when $\gamma < \frac{1}{2}$, although the individual risks are thin-tailed. We define the random variable $\xi = \mathbf{c}^T \mathbf{x}$ and the constant $c = \mathbf{c}^T \mathbf{1}$. We also note that the total portfolio risk changes with the number of investments, N, in a very different way than what is implied by standard diversification results, for which the *size* of the investment is taken for given. If portfolio size were given, then the portfolio risk would vanish as N grew. However, taking into account that the intermediary can change its total exposure with the number of risks, in our framework, the risk does not mitigate, but instead converges to a portfolio risk that is much more heavy-tailed than the individual risks.

Our approach of letting the number of risks within an asset class grow has some similarities with the idea of an *asymptotically fine-grained portfolio* (see Gordy 2000, 2003) used for a theoretical motivation of the capital rules laid out in Basel II. Gordy shows that under the assumptions of one systematic risk factor and infinitely many small idiosyncratic risks, the portfolio invariant VaR rules of Basel II can be motivated. Similar to Gordy's setting, in our model each risk is a small part of the total asset class risk when N is large. However, in our setting all risk is not diversified away when the number of risks increases, as shown in Theorem 2.4.1. In fact, the risk that is not diversified away adds up to systematic risk in our model (which turns into systemic risk if it brings down the whole system), and since the number of risk classes, $M > 1$, there will be multiple risk factors. This is contrary to the analysis in Gordy (2000), where, after diversification, all institutions hold the same one-factor risk. Therefore, the capital rules in Basel II would not be motivated in our model.

2.4.1.2 The Value to Society

We next turn to the value to society of the markets for the M separate risk classes. For the individual intermediary, the game ends if it defaults. From society's perspective, however, we would expect other players to step in and take over the business if an intermediary defaults, although it may take time to set up the business, develop client relationships, etc. Therefore, the cost to society of an intermediary's default may not be as serious as the value lost by the specific intermediary. The argument is that since there is a well-functioning market, it is quite easy to set up a copy of the intermediary that defaulted. We assume that an outside provider of capital steps in and sets up an intermediary identical to the one that defaulted. The argument per assumption, however, only works if there is a well-functioning market. In the unlikely event in which all intermediaries default, the market is not well-functioning and it may take a much longer time to set up the individual copy. Thus, individual intermediary default is less serious from society's perspective, but *massive* intermediary default (for simplicity, the case when all M intermediaries default at the same time) is much more serious. A similar argument is made in Acharya (2009).

To incorporate this reasoning into a tractable model, we use a special numerical mechanism. When only one individual intermediary fails (or a few intermediaries), the market continues to operate reasonably well. In such cases, the replacement of a failed intermediary occurs immediately if an intermediary defaults in an even period ($t = 0, 2, 4, \ldots$), and takes just one period if it defaults in an odd period ($t = 1, 3, 5, \ldots$). We note that in this favorable case without massive default, it takes, on average, half a period to rebuild after an intermediary's default (zero periods or one period with 50% chance each). On the other hand, when all intermediaries fail—a massive default—we assume there is a 50% chance that it will take another $2T$ periods ($2T + 1 - 1$) to return to normal operations. Therefore, the expected extra time to recover after a massive default is $(50\%) \times (2T) + (50\%) \times (0) = T$. Henceforth, we call T the *recovery time after massive default*.

Our distinction between defaults in odd and even time periods significantly simplifies the analysis by ensuring that in even periods there are either 0 or M intermediaries in the market. Without the assumption, we would need to introduce a state variable describing how many intermediaries are alive at each point in time, which would increase the complexity of the analysis. Qualitatively similar results arise when relaxing the assumption, i.e., when instead assuming that it always takes one period to replace up to $M - 1$ defaulting intermediaries and T periods if all M intermediaries default at the same time.

The assumption of longer recovery times after massive intermediary defaults can be viewed as a reduced-form description of the externalities imposed on society by intermediary defaults. The simplicity of the assumption allows us to carry out a rigorous analysis of the role of risk distributions, which is the focus of this section of the book. Micro foundations for such externalities have been suggested elsewhere in the literature, e.g., liquidity risk as in Allen and Gale (2000). In Allen and Gale (2000), the externality occurs when failure of intermediaries triggers failure of other intermediaries.

Society's and individual intermediaries' objectives may be unaligned. From our previous discussion, it follows that the value to society of all intermediaries at $t = 0$ is, in recursive form:

$$V^S = Mdc + \delta qMdc + \delta^2(1 - (1-q)^M)V^S + \delta^{2T+2}(1-q)^M V^S, \quad (2.25)$$

implying that

$$V^S = \frac{Mdc(1 + \delta q)}{1 - \delta^2 + \delta^2(1 - \delta^{2T})(1-q)^M}. \quad (2.26)$$

The first term on the right-hand side of (2.25) is the value generated between $t = 0$ and $t = 1$, and the second term is the expected value between $t = 1$ and $t = 2$.

The third term is the discounted contribution to the value from $t = 2$ and forward if all M intermediaries do not default (which occurs with probability $(1 - (1 - q)^M)$), and the fourth term is the contribution to the value if all M intermediaries default (which occurs with probability $(1 - q)^M$). We shall see that this externality of massive intermediary default in the form of longer recovery time significantly decreases—or even reverses—the value of diversification from society's perspective and that the situations in which diversification is optimal versus suboptimal are easily characterized.

2.4.1.3 Risk-Sharing

We next analyze what happens when the M different intermediaries, with identically distributed risk portfolios, $\xi_1, \ldots, \xi_M$, get the opportunity to share risks. We recall that the risk portfolios in the different risk classes are independently distributed. Therefore, as long as the intermediaries do not trade, a shock to one intermediary will not spread to others. If the intermediaries trade, however, systemic risk may arise when their portfolios become more similar because of trades. In what follows, we explore this idea in detail.

We assume that the intermediaries may trade risks at $t = \frac{1}{2}, \frac{3}{2}, \frac{5}{2}, \ldots$, after they have formed their portfolio, but before the risks are realized. We assume that the VaR requirement must hold at all times. For example, even if the intermediaries trade with each other at $t = \frac{1}{2}$, the VaR requirement needs to be satisfied between $t = 0$ and $t = \frac{1}{2}$.[41] Also, the counter-party of the risk trade correctly anticipates whether trades between intermediaries will take place at $t = \frac{1}{2}$, taking the impact of such a trade on the default option, Q, into account when the price is decided.

We focus on symmetric equilibria, in which all intermediaries choose to share risks fully. We define

$$q_M = P\left(\frac{-\sum_{m=1}^M \xi_m}{M} \leq K \right),$$

so q_M is the probability that total losses in the market are lower than total capital. Thus, $1 - q_M$, is the probability for massive intermediary default if intermediaries fully share risks. We note, in passing, that systemic risk is generated through a different mechanism in our model than in Acharya (2009). In Acharya (2009), systemic risk arises when intermediaries take on correlated real investments, and intermediaries do not trade risks. In our model, the risk portfolios of different intermediaries are independent, and systemic risk arises when intermediaries become

---

[41]This is not a critical assumption. Alternatively, we could have assumed that the regulator anticipates whether the individual intermediaries will trade risks, and adjusts the VaR requirements accordingly.

interdependent through trades. The latter type of systemic risk may have been especially important in the financial crisis.

All intermediaries hold portfolios of the same risk features, albeit in different risk classes, so when they share risks, the net "price" they pay at $t = \frac{1}{2}$ is zero. However, the states of the world in which default occurs are different when the intermediaries share risks. In fact, defaults will be perfectly correlated in the full risk-sharing situation: With probability $1 - q_M$ a massive intermediary default occurs and with probability q_M no intermediary defaults. Risk-sharing could, of course, be achieved not only by cross-ownership, but also by trading in derivatives contracts (credit default swaps (CDS), corporate debt, etc.).

For individual intermediaries, the value when sharing risks (using the same arguments as when deriving (2.18)) is then

$$V_M^B = \frac{dc}{1 - \delta q_M}, \tag{2.27}$$

which should be compared with the value of not sharing, (2.18). Therefore, as long as

$$q_M > q, \tag{2.28}$$

the intermediaries prefer to trade risks, since it leads to a higher probability of survival, and thereby a higher value.

We note that there may be additional reasons for intermediaries to prefer risk-sharing. For example, if intermediaries anticipate that they will be bailed out in case of massive defaults, if managers are less punished if an intermediary performs poorly when all other intermediaries perform poorly too, or if the VaR restrictions are relaxed, this provides additional incentives for risk-sharing, which would amplify our main result.

2.4.1.4 The Value to Society

From society's perspective, a similar argument as when deriving Eqs. (2.25) and (2.26) shows that

$$V_M^S = Mdc + \delta q_M Mdc + \delta^2 q_M V_M^S + \delta^{2T+2}(1 - q_M)V_M^S, \tag{2.29}$$

so the total value of risk-sharing between intermediaries is

$$V_M^S = \frac{Mdc(1 + \delta q_M)}{1 - \delta^2 + \delta^2(1 - \delta^{2T})(1 - q_M)}. \tag{2.30}$$

Therefore, via (2.26), it follows that society prefers risk-sharing when

$$\frac{1 + \delta q_M}{1 + \lambda(1 - q_M)} > \frac{1 + \delta q}{1 + \lambda(1 - q)^M}, \tag{2.31}$$

where

$$\lambda = \frac{\delta^2(1 - \delta^{2T})}{1 - \delta^2}. \tag{2.32}$$

The value of λ determines the relative trade-off society makes between the costs in foregone investment opportunities of individual and massive intermediary default.[42] If $T = 0$, there is no extra delay when massive default occurs. In this case, $\lambda = 0$, and society's trade-off (2.31) is the same as individual intermediaries' (2.28). If, on the other hand, T is large and δ is close to one, representing a situation in which it takes a long while to set up a market after massive default and the discount rate is low, λ is large, and the trade-off between $(1 - q_M)$ and $(1 - q)^M$ in (2.31) becomes very important. In this case, society will mainly be interested in minimizing the risk of massive default and this risk is minimized when intermediaries do *not* share risk, since $1 - q_M \geq (1 - q)^M$.[43]

It is clear that society and individual intermediaries may disagree about whether it is optimal to diversify. Specifically, this occurs when (2.28) holds, but (2.31) fails. The opposite, that (2.28) fails but (2.31) holds, can never occur.[44] To understand the situation conceptually, we study the special case when there are two risk classes, $M = 2$. In Fig. 2.4, we show the different outcomes depending on the realizations of losses in the diversified and separated cases. When both $-\xi_1 > K$ and $-\xi_2 > K$, it does not matter whether the intermediaries diversify or not, since there will be massive intermediary default either way. Similarly, if $-\xi_1 \leq K$ and $-\xi_2 \leq K$, neither intermediary defaults, regardless of whether they diversify or not. However, in the case when $-\xi_1 > K$ and $-\xi_2 \leq K$, the outcome is different, depending on the diversification strategy the intermediaries have chosen. In this case, if $-\xi_1 - \xi_2 > 2K$, then a massive default occurs if the intermediaries are diversified, but only a single default if they are not. If $-\xi_1 - \xi_2 \leq 2K$, on the other hand, no default occurs if the intermediaries are diversified, but a single default occurs if they are not. Exactly the same argument applies in the case when $-\xi_1 \leq K$ and $-\xi_2 > K$. Therefore, the optimal outcome depends on the trade-off between avoiding single defaults in some states of the world but introducing massive defaults in others, when diversifying. One should note that the point that risk-sharing increases the risk for joint failure was originally made in Shaffer (1994).

---

[42]It is easy to show that $\lambda \in [0, T]$, and that λ is increasing in δ and T.

[43]This follows trivially, since $1 - q_M = P(\sum_i -\xi_i > MK) \geq P(\cap_i\{-\xi_i > K\}) = (1 - q)^M$.

[44]This also follows trivially, since $1 - q_M \geq (1 - q)^M$ and therefore, if $q \geq q_M$, then $\frac{1+\delta q}{1+\lambda(1-q)^M} \geq \frac{1+\delta q}{1+\lambda(1-q_M)} \geq \frac{1+\delta q_M}{1+\lambda(1-q)^M}$.

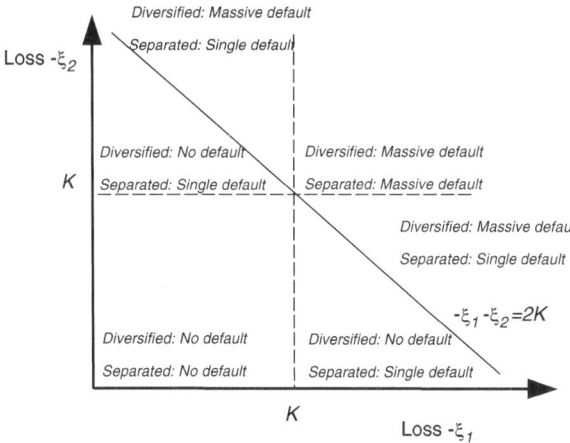

Fig. 2.4 Cost to society for diversified and separated cases, when there are two intermediaries. In the diversified case, massive default occurs if $-\xi_1 - \xi_2 \geq 2K$. In the separated case, massive default occurs if $-\xi_1 \geq K$ and $-\xi_2 \geq K$, whereas single default occurs if $-\xi_1 \geq K$ and $-\xi_2 < K$, or if $-\xi_2 \geq K$ and $-\xi_1 < K$. Thus, massive default is rarer in the separated case than in the diversified case, but it is more common that at least one intermediary defaults in the separated case

It is clear that the objectives of the intermediaries and society will depend on the distributions of the ξ-risks and it may not be surprising that standard results from the theory of diversification apply to the individual intermediaries' problem. Society's objective function is more complex, however, since it trades off the costs of massive and individual intermediary defaults. It comes as a pleasant surprise that for low default risks (i.e., for a β close to zero in the VaR$_\beta$ constraint) given the distribution of the ξ-risk, we can completely characterize when the objectives of intermediaries and society are different.

Theorem 2.4.2 *Let $\xi_1, \ldots, \xi_M$ be i.i.d. asset class risks that follow power law (1.2) in the left tails with the tail index $\zeta \neq 1$, and d, δ, and T as previously defined. Then, for low β,*

a) From an intermediary's perspective, risk-sharing is optimal if and only if $\zeta > 1$, regardless of the number of risk classes, M.

b) From society's perspective, risk-sharing is optimal if and only if $\zeta > 1$ and $M > M_*$, where M_* is the diversification threshold

$$M_* = \left(\frac{1 - \delta^{2T+1}}{1 - \delta} \right)^{\frac{1}{\zeta - 1}}. \tag{2.33}$$

Thus, the break-even for individual intermediaries, from (2.28), is a power law distribution with the tail index of one, e.g., a Cauchy distribution. If the distribution is heavier, then (2.28) fails, and intermediaries and society therefore agree that there should be no risk-sharing. For thinner tails, intermediaries prefer risk-sharing. On the contrary, for society, what is optimal depends on M. It is easy to show that $M_* \in [1, (2T + 1)^{\frac{1}{\zeta-1}})$, and that M_* is increasing in δ and T. We note that for the limit case when $\delta \to 0$, $\zeta \to \infty$, or $T = 0$, i.e., when $M_* \to 1$, then society and intermediaries always agree.

It is useful to define

$$\eta = \frac{1 - \delta^{2T+1}}{1 - \delta}, \tag{2.34}$$

so that $M_* = \eta^{\frac{1}{\zeta-1}}$. This separates the impact on M_* of the tail behavior of the risk distributions from the other factors (δ and T). It follows immediately that $\eta \in [1, 2T + 1)$.

From Theorem 2.4.2, it follows that if there is a large number of risk classes, M, society agrees with the individual intermediaries:

Corollary 2.4.1 *Given i.i.d. risk classes that follow power laws (1.2) with the tail index ζ, if there is a large enough number of risk classes available, intermediaries and society agree on whether risk-sharing is optimal.*

In Fig. 2.5, we show the break-even number of risk classes, M, for heavy-tailed risks with the tail indices $\zeta = 2$, 2.5, 3, 4. With no externality of massive default ($\eta = 1$), society prefers diversification when $\zeta > 1$, just like individual intermediaries. However, when $\eta > 1$, diversification is suboptimal up until M_*. For example, for $\zeta = 2$, and $\eta = 10$, at least ten risk-classes are needed for diversification to be optimal for society.

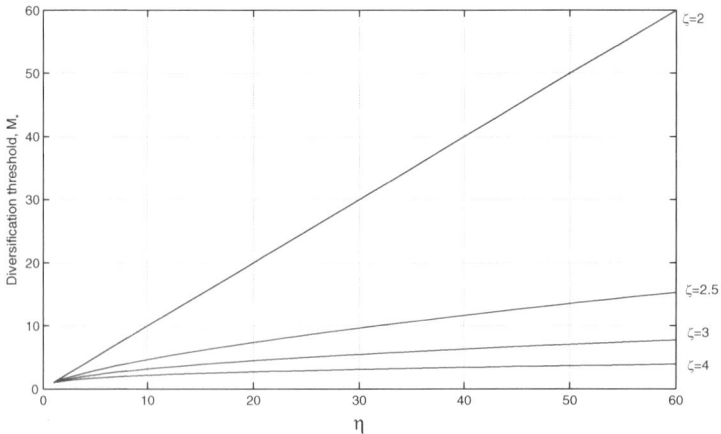

Fig. 2.5 Diversification threshold, M_*, for heavy-tailed risk distributions with tail indices $\zeta = 2, 2.5, 3, 4$, as a function of the coefficient $\eta = \frac{1-\delta^{2T+1}}{1-\delta}$, where δ is the per period discount factor and T is the recovery time after massive default. The diversification threshold is the break-even number of risk classes needed for diversification to be optimal for society. Results are asymptotically valid for VaR loss probabilities β close to zero in (2.22)

Theorem 2.4.2 provides a complete characterization of the objectives of intermediaries and firms for VaR loss probabilities close to zero, by relating the number of risk classes, M, the discount rate, δ, the tail index of the risks, ζ, and the recovery time after massive default, T. It also relates to the uncertainty of correlations, γ, through the relation $\zeta = 2\gamma$. Theorem 2.4.2 is therefore the main result of this section. The theorem has several immediate empirical implications, since when (2.33) fails, we would expect there to exist financial regulations against risk-sharing across risk classes:

Implication 2.4.1 *Economies with heavier-tailed risk distributions should have stricter regulations for risk-sharing between risk classes.*

Equivalently, using the relationship between uncertainty of correlation structure and tail distributions,

Implication 2.4.2 *Economies with more uncertain correlation between risk classes should have stricter regulations for risk-sharing.*

We also have

Implication 2.4.3 *Economies with lower interest rates should have stricter regulations for risk-sharing.*

Implication 2.4.4 *Economies with fewer risk classes should have stricter regulations for risk-sharing.*

Implication 2.4.5 *Economies with risk classes for which it takes longer to recover after a massive default should have stricter regulations for risk-sharing.*

Implication 2.4.4 may be related to the size of the economy, in that larger economies may have more risk classes and therefore, society may be more tolerant to risk-sharing across these classes. Moreover, Implication 2.4.5 may be related to the degree to which an economy is open to foreign investments in that economies that are open may be faster in recovering after a massive default, and thereby allow more risk-sharing between risk classes. In Sect. 2.4.2, we apply these implications to derive policy solutions for controlling the social costs created when actions by intermediaries to diversify create investments in highly correlated market portfolios.

2.4.2 Potential Implications for Risk Management and for Policy Makers

Capital requirements provide the most common mechanism used to control the risk of bank failure. Set at a high enough level, a simple capital-to-asset requirement can achieve any desired level of safety for an individual bank. Such capital requirements, however, impose significant costs on banks by limiting their use of debt tax shields, expanding the problem of debt overhang, and creating agency problems for the shareholders.[45] These costs have a negative impact on the overall economy because they reduce the efficiency of financial intermediation.[46] In our model, it is clear that increasing capital requirements is an imperfect tool for the regulator, since it cannot be used to specifically target negative externalities of systemic risk. In fact, there is a one-to-one relationship between VaR and capital requirements in our model—increasing the capital requirement is equivalent to decreasing the VaR.

For this reason, new proposals to control systemic risk in the banking sector recommend focused capital requirements based on each bank's specific contribution to the aggregate systemic risk. Acharya (2009), for example, advocates higher capital requirements for banks holding asset portfolios that are highly correlated with the portfolios of other banks. This follows from his model in which banks create systemic risk by choosing correlated portfolios. In a similar spirit, Adrian and Brunnermeier (2011) advocate a "CoVaR" method in which banks face higher capital requirements based on their measured contribution to the aggregate systemic risk.

Focused capital requirements will be efficient in controlling systemic risk, however, only if the source of the systemic risk is properly identified. In particular, the model in this section creates a symmetric equilibrium in which each of the M banks is responsible for precisely $1/M$ of the systemic risk. Furthermore, systemic risk in our model arises only as a by-product—a true negative externality—of each bank's attempt to eliminate its own idiosyncratic risk. The risks that they take on are independent, but the diversification changes the states of the world in which massive default occurs. For this reason, neither VaR constraints, nor the capital requirement plans advocated by Acharya (2009) will be effective in controlling the type of systemic risk that arises in our model. The CoVaR measure in Adrian and Brunnermeier (2011) may also be imperfect, since it does not take tail-distributions into account. Of course, systemic risk in the banking industry may reflect a variety of generating mechanisms, so we are not claiming anything like a monopoly on the proper regulatory response. But it is the case that even focused capital requirements will not be effective if the systemic risk arises because banks

---

[45]The debt overhang problem arises in recapitalizing a bank because the existing shareholder ownership is diluted while some of the cash inflow benefit accrues as a credit upgrade for the existing bondholders and other bank creditors. The agency problems arise because larger capital ratios provide management greater incentive to carry out risky investments that raise the expected value of compensation but may reduce expected equity returns; for further discussion, see Kashyap et al. (2008). While the tax shield benefit of debt is valuable for the banking industry, it is not necessarily welfare-enhancing for society.

[46]The efficiency costs of capital requirements can be mitigated by setting the requirements in terms of contingent capital in lieu of balance sheet capital. One mechanism is based on bonds that convert to capital if bankruptcy is threatened (Flannery 2005), but that instrument is not particularly directed to systemic risk. Kashyap et al. (2008) take the contingent capital idea a step further by requiring banks to purchase "capital insurance" that provides cash to the bank if industry losses, or some comparable aggregate trigger, hits a specified threshold. This mechanism may reduce or eliminate the costs that are otherwise created by bankruptcy, but it does not eliminate the negative externality that creates the systemic risk.

hedge their idiosyncratic risk by swapping into a "market" portfolio that is then held in common by all banks.

For the case developed in our model, where all banks wish to adopt the same diversified market portfolio, direct prohibitions against specific banking activities or investments in specific asset classes will be more effective than capital requirements as a mechanism to control systemic risk. For example, Volcker (2010) has recently proposed to restrict commercial banking organizations from certain proprietary and more speculative activities. While such prohibitions may seem draconian, they would apply only to activities or asset classes in which moderately heavy tails create a discrepancy between the private and public benefits of diversification. No regulatory action would be needed for asset classes with thin-tailed risks, where the banks and society both benefit from diversification, and for severely heavy-tailed risks, where the banks and society agree that diversification is not beneficial.

It is also useful to note that direct prohibitions have long existed in U.S. banking regulation. For one thing, U.S. commercial banks have long been prohibited from investing in equity shares. Even more relevant, the 1933 Glass Steagall Act forced U.S. commercial banks to divest their investment banking divisions. Subsequent legislation—specifically the 1956 Bank Holding Company Act and the Gramm–Leach–Bliley Act of 1999—provided more flexibility, by expanding the range of allowed activities for a bank holding company, although commercial banks are themselves still restricted to a "banking business." Glass–Steagall thus provides a precedent for direct prohibitions on bank activities as well as an indication that the prohibitions can be changed over time as conditions warrant.

Experience with regulating catastrophe insurance counter-party risk suggests another practical and specific regulatory approach, namely "monoline" requirements. Monoline requirements have long been successfully imposed on insurance firms that provide coverage against default by mortgage and municipal bond borrowers; see Jaffee (2006b, 2009). The monoline requirements prohibit these insurers from operating as multiline insurers that offer coverage on multiple insurance lines. The monoline restriction eliminates the possibility that large losses on the catastrophe line would bankrupt a multiline insurer, thus creating a cascade of losses for policyholders across its other insurance lines. Such monoline restrictions do create a cost in the form of the lost benefits of diversification, because a monoline insurer is unable to deploy its capital to pay claims against a portfolio of insurance risks. Nevertheless, Ibragimov et al. (2012) show that when the benefits of diversification are muted by heavy tails or other distributional features, the social benefits of controlling the systemic risk dominate the lost benefits of diversification.

As a specific example, we reference the use of CDS purchased by banks and other investors to provide protection against default on the subprime mortgage securities they held in their portfolios. The systemic problem was that the CDS protection was provided by other banks and financial service firms acting as banks, i.e., American International Group, Inc. (AIG), with the effect that a set of large banks ended up holding a very similar, albeit diversified, portfolio of subprime mortgage risks.[47] When the risks on the individual underlying mortgages proved to be highly correlated, this portfolio suffered enormous losses, creating the systemic crisis. Capital requirements actually provided the investing banks with incentive to purchase the CDS protection, so that higher capital requirements, per se, do not solve the systemic problem. Instead, there must be regulatory recognition that moderately heavy-tailed risk distributions create situations in which the social costs may exceed the private benefits of diversification.

2.5 Marketing: Optimal Bundling Under Heavy-Tailed Valuations[48]

In December 2007, two tickets to a Led Zeppelin reunion concert in London were sold in a charity auction. The face value of the two tickets was in total GBP 250 and when the winning bid turned out to be GBP 83,000—332 times the face value—this made headlines across the world. Kenneth Donnell, 25, who bought the tickets stated in interviews that he had wanted to see his father's (sic!) favorite band live for years and that he had been sober when he joined the auction. Although the demand for tickets to the concert had been overwhelmingly higher than the supply, it is plausible to assume that most buyers' valuations of the tickets were far lower than what Mr. Donnell paid.

The case of Mr. Donnell and the Led Zeppelin tickets is just one example of very diverse private valuations observed in markets for cultural and sport events, as well as in those for antiques and collectibles and online auctions and marketplaces such as eBay and StubHub. In these markets, bundling of goods is common practice and a natural question is then what the consumers' and the seller's preferences over bundles are when the buyers have diverse private valuations.

The problem of optimal bundling strategy has received much attention over the last quarter of a century in the marketing and economics literature (see, e.g., the review in Stremersch and Tellis 2002, and the references therein). However, the importance of the distribution of consumer valuations has, to the best of our knowledge, not been emphasized.

In this section we analyze the optimal bundling strategy for a multiproduct monopolist when the distribution of consumer valuations is *heavy-tailed*. We do this for two situations. In the first, the seller chooses how to bundle a given set of goods and sell the bundles in different auctions. In the second, he/she produces and

---

[47]It is important to note that AIG wrote its CDS contracts from its Financial Products subsidiary, which was chartered as a savings and loan association and not as an insurance firm. Indeed, AIG also owns a monoline mortgage insurer, United Guaranty, but this subsidiary was not the source of the losses that forced the government bailout.

[48]This section has drawn upon material from the article Ibragimov and Walden (2010), which was published in the *Management Science*, Vol. 56, Issue 11, pp. 1963–1976.

provides the bundles for profit-maximizing prices. We focus on the analysis of pure bundling with one set of bundles offered for sale, as opposed to mixed bundling, in which consumers can choose among all possible sets of bundles (see Adams and Yellen 1976; McAfee and Whinston 1989).

In the auction case, our main contribution is to complement and generalize the previous literature, e.g., Palfrey (1983), to the case of heavy-tailed valuations. Palfrey (1983) showed that in the case of two buyers, the seller will prefer to bundle the products. The two buyers, in contrast, unanimously prefer separate auctions to any other bundling decision. Palfrey (1983) further showed that, with bounded valuations, if there are more than two buyers, they will *never* unanimously prefer separate auctioning of the goods. This section demonstrates that, on the contrary, with extremely heavy-tailed distributions, the buyers *always* unanimously prefer separate auctioning. The key distinction between the main results in Palfrey (1983) and ours is the distributional assumption on consumers' valuations.

In the case of profit-maximizing prices, the results of previous literature are completely reversed when valuations are extremely heavy-tailed. For instance, the results in Bakos and Brynjolfsson (1999) and Fang and Norman (2006) indicate that, with thin-tailed valuations such as those with log-concave distributions, the optimal strategy for a multiproduct monopolist is to bundle goods with low marginal costs and to separately sell products with high marginal costs. We show that, to the contrary, under extremely heavy-tailed valuations the monopolist prefers bundling goods with high marginal costs and separately providing goods with low marginal costs. However, the results in the thin-tailed case in the previous literature continue to hold for moderately heavy-tailed valuations.

The main reason why the results are so different under heavy-tailed valuations is the following. Under *thin-tailed* valuations, consumers' valuations per good for a bundle typically have a lower spread, measured by variance, relative to the valuations for individual goods (see the discussion in Bakos and Brynjolfsson 1999; Fang and Norman 2006; Palfrey 1983; Salinger 1995; Schmalensee 1984). Similarly, under *moderately* heavy-tailed reservation prices, the consumers' valuations per good for bundles have less spread relative to the valuations for component products, as measured by their *peakedness*.[49] Under *extremely* heavy-tailed valuations, this property is reversed: in this case, the spread of reservation prices per product for bundles, as measured by peakedness, is greater than that of valuations for components.[50] In the auction setting, given the relatively high spread of valuations of the bundled goods, the potential upside for the seller is then very high. Therefore, since the actual price is based on the second highest bid, the most important thing for the seller is to increase the chances that multiple buyers with high valuation bid in the same auction, which is achieved by bundling. The argument is reversed for the

---

[49]The terms "reservation prices" and "valuations" are used as synonyms in this section, in accordance with the well-established tradition in the bundling literature.

[50]The arguments for the results in this section are based on peakedness, majorization and VaR comparison results for heavy-tailed distributions discussed in Sects. 2.1.3 and 2.1.4.

buyers. Similar arguments can be made for the results with a monopolist producer, as elaborated upon in the section.

The rest of the section is organized as follows. In Sect. 2.5.1, we discuss related literature. Sect. 2.5.2 reviews the framework for modeling optimal bundling. Sections 2.5.3 and 2.5.4 contain our main results in the section, which, for tractability, are given in a rather special setting with i.i.d. valuations and heavy-tailed stable distributions discussed in Sect. 2.1.2.

2.5.1 Related Literature

Many studies have emphasized that bundling decisions of a monopolist providing two goods depend on correlations between consumers' valuations for the products (see Adams and Yellen 1976; McAfee and Whinston 1989; Salinger 1995; Schmalensee 1984), the degrees of complementarity and substitutability between the goods (e.g., Dansby and Conrad 1984; Lewbel 1985; Venkatesh and Kamakura 2003) and the marginal costs for the products (see, among others, Salinger 1995; Venkatesh and Kamakura 2003).

Most of these studies on bundling have focused, however, on prescribed distributions for valuations in the case of two products and their packages, such as bivariate uniform or Gaussian distributions, and only a few general results are available for larger bundles (e.g., Bakos and Brynjolfsson 1999, 2000a,b; Chu et al. 2011; Fang and Norman 2006; Palfrey 1983). For instance, Palfrey (1983) obtained characterizations of the monopolist's and buyers' preferences over bundled Vickrey auctions with valuations concentrated on a finite interval. In a related paper, Chakraborty (1999) obtained characterizations of optimal bundling strategies for a monopolist providing two independently priced goods on Vickrey auctions under a regularity condition on quantiles of bidders' valuations. As follows from Proschan (1965) results discussed in Remark 2.1.3 in Sect. 2.1.3, this regularity condition is satisfied for symmetric valuations with log-concave densities.[51]

Bakos and Brynjolfsson (1999) investigated optimal bundling decisions for a multiproduct monopolist providing large bundles of independently priced goods with zero marginal costs (information goods) for profit-maximizing prices to consumers whose valuations belong to a class that includes, again by Proschan (1965), reservation prices with log-concave densities symmetric about the mean.[52]

---

[51]From Theorems 2.1.1 and 2.1.2 it further follows that the regularity condition is also satisfied for moderately heavy-tailed valuations, but it does not hold for extremely heavy-tailed valuations. Therefore, Chakraborty's analysis cannot be applied if consumers' valuations are extremely heavy-tailed.

[52]In particular, the assumptions are satisfied for valuations with a finite support $[\underline{v}, \overline{v}]$ distributed as the truncation $XI(|X - \mu| < h)$, $h > 0$, of an arbitrary random variable X with a log-concave density symmetric about $\mu = (\underline{v} + \overline{v})/2$, where $h = (\overline{v} - \underline{v})/2$ and $I(\cdot)$ is the indicator function (see also Remark 2 in An 1998).

Among other results, Bakos and Brynjolfsson (1999) showed that, for this class of valuations, if the seller prefers bundling a certain number of goods to selling them separately and if the optimal price per good for the bundle is less than the mean valuation, then bundling any greater number of goods will further increase the seller's profits, compared to the case where the additional goods are sold separately. According to the result, in the above settings, a form of superadditivity for bundling decisions holds, that is, the benefits to the seller grow as the number of goods in the bundle increases.

Fang and Norman (2006) showed that a multiproduct monopolist providing bundles of independently priced goods to consumers with valuations with log-concave densities prefers selling them separately to any other bundling decision if the marginal costs of all the products are greater than the mean valuation; under some additional distributional assumptions, the seller prefers providing the goods as a single bundle to any other bundling decision if the marginal costs of the goods are identical and are less than the mean reservation price.

Chu et al. (2011) focus on the analysis of near optimality of bundle-size pricing where the prices for bundles depend (only) on their size. They also provide a range of numerical experiments for different cost scenarios and distributional assumptions on consumers' valuations, including exponential, logit, uniform, multivariate normal and multivariate lognormal distributions, and an empirical analysis of pricing schemes for a theater company offering tickets for eight different plays or musicals and their packages.

Hitt and Chen (2005) and Wu et al. (2008) have focused on the analysis of customized bundling of information goods, a pricing strategy under which consumers can choose a certain quantity of goods sold for a fixed price. The results in these papers, in particular, show that, under some commonly used assumptions, the mixed-bundle problem can be reduced to customized bundling. They further demonstrate how the customized-bundle solution is affected by heterogeneity and correlations in customers' valuations and by complementarity or substitutability among the goods sold.

2.5.2 A Framework for Modeling Optimal Bundling

We consider a setting with a single seller providing m goods to n consumers. Let $M = \{1, 2, \ldots, m\}$ be the set of goods sold on the market and let $J = \{1, 2, \ldots, n\}$ denote the set of buyers. Let 2^M stand for the set of all subsets of M. As in Palfrey (1983), the seller's bundling decisions $\mathcal{B}$ are defined as partitions of the set of items M into a set of subsets, $\{B_1, \ldots, B_l\} = \mathcal{B}$; the subsets $B_s \in 2^M$, $s = 1, \ldots, l$, are referred to as bundles. That is, $B_s \neq \emptyset$ for $s = 1, \ldots, l$; $B_s \cap B_t = \emptyset$ for $s \neq t$, $s, t = 1, \ldots, l$; and $\cup_{s=1}^{l} B_s = M$ (see Bakos and Brynjolfsson 1999; Fang and Norman 2006; Palfrey 1983). It is assumed that the seller can offer one (and only one) partition $\mathcal{B}$ for sale on the market (this is referred to as pure bundling, see Adams and Yellen 1976). We denote by $\underline{\mathcal{B}} = \{\{1\}, \{2\}, \ldots, \{m\}\}$ and

$\overline{\mathcal{B}} = \{1, 2, \ldots, m\}$ the bundling decisions corresponding, respectively, to the cases where the goods are sold separately (that is, on separate auctions or using unbundled sales) and as a single bundle M.

For a bundle $B \in 2^M$, we write card(B) for the number of elements in B and denote by π_B the seller's profit resulting from selling the bundle, with the convention that $\pi_B = 0$ if the bundle is not sold. For a bundling decision $\mathcal{B} = \{B_1, \ldots, B_l\}$, we write $\Pi_\mathcal{B}$ for the seller's total profit resulting from following $\mathcal{B}$, that is, $\Pi_\mathcal{B} = \sum_{s=1}^{l} \pi_{B_s}$.

A risk-neutral seller prefers (strictly prefers) a bundling decision $\mathcal{B}_1$ to a bundling decision $\mathcal{B}_2$ *ex ante* if $E\Pi_{\mathcal{B}_1} \geq E\Pi_{\mathcal{B}_2}$ (resp., if $E\Pi_{\mathcal{B}_1} > E\Pi_{\mathcal{B}_2}$), where E denotes the expectation operator. The seller prefers a bundling decision $\mathcal{B}_1$ to a bundling decision $\mathcal{B}_2$ *ex post* if $\Pi_{\mathcal{B}_1} \geq \Pi_{\mathcal{B}_2}$ (a.s.), that is, if $P(\Pi_{\mathcal{B}_1} \geq \Pi_{\mathcal{B}_2}) = 1$. More generally, if the seller has an increasing utility of wealth function $U : \mathbf{R}_+ \to \mathbf{R}$, then she prefers (strictly prefers) a bundling decision $\mathcal{B}_1$ to a bundling decision $\mathcal{B}_2$ if $EU(\Pi_{\mathcal{B}_1}) \geq EU(\Pi_{\mathcal{B}_2})$ (resp., if $EU(\Pi_{\mathcal{B}_1}) > EU(\Pi_{\mathcal{B}_2})$). The setting with a concave function U represents the case of a risk-averse seller. This section focuses on characterizations of the seller and buyers' *ex ante* preferences over bundles of goods sold.

Consumers' preferences over the bundles $B \in 2^M$ are determined by their valuations (reservation prices) $v(B)$ for the bundles and, in particular, by their valuations $X_i = v(\{i\})$ for goods $i \in M$ (when the goods are sold separately) which are referred to as stand-alone valuations. Consumers' valuations for bundles of goods are assumed to be additive in those of component goods:

$$v(B) = \sum_{i \in B} v(\{i\}) = \sum_{i \in B} X_i \tag{2.35}$$

and their utilities from consuming goods in $\mathcal{B} = \{B_1, \ldots, B_l\}$ are given by

$$v(\mathcal{B}) = \sum_{s=1}^{l} v(B_s) = \sum_{s=1}^{l} \sum_{i \in B_s} v(\{i\}) = \sum_{s=1}^{l} \sum_{i \in B_s} X_i = \sum_{i=1}^{m} X_i. \tag{2.36}$$

If additivity conditions (2.35) and (2.36) hold, then the products provided by the monopolist are said to be *independently priced* (see Venkatesh and Kamakura 2003).

In the case where the valuations for bundles are nonnegative: $v(B) \geq 0, B \in 2^M$, it is said that the goods in M and their bundles satisfy the *free disposal* condition. The free disposal condition is particularly important in the case of information goods and in the economics of the Internet (see Bakos and Brynjolfsson 1999, 2000a,b). In Sect. 2.5.4, the valuations $v(B)$ are allowed to be negative. This corresponds to the situation where the goods have negative value to some consumers (e.g., articles exposing certain political views, advertisements or pornography in the case of information goods, see Bakos and Brynjolfsson 1999).

For our main results presented in the next two sections, X_i, $i \in M$, denote i.i.d. r.v.'s representing the distribution of consumers' valuations for goods $i \in M$ that determine their reservation prices for bundles.

For $j \in J$, the jth consumer's valuations for goods in M are assumed to be $\tilde{X}_{ij}$, $i \in M$, where $\tilde{X}^{(j)} = (\tilde{X}_{1j}, \ldots, \tilde{X}_{mj})$, $j \in J$, are independent copies of the vector $(X_1, \ldots, X_m)$, and her reservation prices $v_j(B)$ for bundles $B \in 2^M$ of goods in M are given by $v_j(B) = \sum_{i \in B} \tilde{X}_{ij}$. The seller is assumed to know only the distribution of consumers' reservation prices for goods in M and their bundles. The valuations $v_j(B)$ for bundles $B \in 2^M$, are known to buyer j; however, the buyer has only the same incomplete information about the other consumers' reservation prices as does the seller (see Palfrey 1983).

2.5.3 Optimal Bundled Auctions with Heavy-Tailed Valuations

Let us first consider the case in which the goods in M and their bundles are provided by a seller through Vickrey auctions (see Palfrey 1983). The Vickrey auctions are separate and independent, one per bundle. In this setting, the buyers submit simultaneous sealed bids for bundles of goods. The highest bid wins the auction and pays the seller the second highest bid. It is well-known that, in such a setup, under additivity conditions (2.35) and (2.36), a dominant strategy for each bidder is to bid her true valuations for goods and their bundles.

Let $j \in J$ and let $\tilde{x}^{(j)} = (\tilde{x}_{1j}, \ldots, \tilde{x}_{mj}) \in \mathbf{R}^m_+$. If a bundle B consisting of independently priced goods is offered for sale in a Vickrey auction, then the expected surplus, $S_j(B, \tilde{x}^{(j)})$, to consumer j with the values of stand-alone valuations $\tilde{X}^{(j)} = \tilde{x}^{(j)}$ and induced valuations for bundles $v_j(B) = \sum_{i \in B} \tilde{x}_{ij}$, $B \in 2^M$ (the conditional expectation of the surplus to the consumer, conditioning on $\tilde{X}^{(j)} = \tilde{x}^{(j)}$) is

$$ES_j(B, \tilde{x}^{(j)}) =$$

$$P\left(\max_{s \in J, s \neq j} v_s(B) < v_j(B) \right)\left(v_j(B) - E\left(\max_{s \in J, s \neq j} v_s(B) \,\Big|\, \max_{s \in J, s \neq j} v_s(B) < v_j(B) \right) \right),$$

where $v_s(B) = \sum_{i \in B} \tilde{X}_{is}$, $B \in 2^M$, $s \in J$, $s \neq j$ (see Palfrey 1983). If the seller follows a bundling decision $\mathcal{B} = \{B_1, \ldots, B_l\}$, then the expected surplus $S_j(\mathcal{B}, \tilde{x}^{(j)})$ to the jth buyer with the vector of stand-alone valuations $\tilde{X}^{(j)} = \tilde{x}^{(j)}$ is $ES_j(\mathcal{B}, \tilde{x}^{(j)}) = \sum_{s=1}^{l} ES_j(B_s, \tilde{x}^{(j)})$. The jth buyer with $\tilde{X}^{(j)} = \tilde{x}^{(j)}$ is said to prefer (strictly prefer) a bundling decision $\mathcal{B}_1$ to a bundling decision $\mathcal{B}_2$, *ex ante*, if $ES_j(\mathcal{B}_1, \tilde{x}^{(j)}) \geq ES_j(\mathcal{B}_2, \tilde{x}^{(j)})$ (resp., if $ES_j(\mathcal{B}_1, \tilde{x}^{(j)}) > ES_j(\mathcal{B}_2, \tilde{x}^{(j)})$). If all buyers $j \in J$ (strictly) prefer a bundling decision $\mathcal{B}_1$ to a bundling decision $\mathcal{B}_2$ *ex ante* for *almost all* realizations of their valuations $\tilde{X}^{(j)}$, it is said that buyers *unanimously* (strictly) prefer $\mathcal{B}_1$ to $\mathcal{B}_2$ *ex ante*. More precisely, buyers unanimously prefer (strictly prefer) a partition $\mathcal{B}_1$ to a partition $\mathcal{B}_2$ if, for all $j \in J$, $P[E(S_j(\mathcal{B}_1, \tilde{X}^{(j)})|\tilde{X}^{(j)}) \geq$

$E(S_j(\mathcal{B}_2, \tilde{X}^{(j)})|\tilde{X}^{(j)})] = 1$ (resp., $P[E(S_j(\mathcal{B}_1, \tilde{X}^{(j)})|\tilde{X}^{(j)}) > ES_j(\mathcal{B}_2, \tilde{X}^{(j)})|\tilde{X}^{(j)})] = 1$), where, as usual, $E(\cdot|\tilde{X}^{(j)})$ stands for the expectation conditional on $\tilde{X}^{(j)}$.[53]

In accordance with the assumption of nonnegativity of bids and consumers' valuations usually imposed in the auction theory, we focus on the case where consumers' valuations for goods and bundles provided are nonnegative and model them using the framework of positive (extremely heavy-tailed) stable r.v.'s. To simplify the presentation, we consider the case of positive extremely heavy-tailed valuations with Lévy densities (2.2) in Sect. 2.1.2 and the tail index and the index of stability $\zeta = \alpha = 0.5$. The results continue to hold for all other extremely heavy-tailed positive stable valuations with stable distributions with the indices of stability $\alpha < 1$ and the skewness parameter $\beta = 1$ (in notation of Sect. 2.1.2, the distributions $S_\alpha(\sigma, 1, \mu)$ with $\alpha < 1$ and $\mu \geq 0$ that are concentrated on the semi-axis $[\mu, \infty)$). In addition, they continue to hold for symmetric extremely heavy-tailed valuations from the class $\underline{\mathcal{CS}}$.

Theorem 2.5.1 shows that consumers unanimously prefer (*ex ante*) separate provision of goods on Vickrey auctions to any other bundling decision in the case of an arbitrary number of buyers, if their valuations are extremely heavy-tailed. In the case of more than two buyers, these results are reversals of those given by Theorem 6 in Palfrey (1983) from which it follows that if consumers' valuations are concentrated on a finite interval, then the buyers never unanimously prefer separate provision auctions (Theorem 2.5.1 does not contradict Theorem 6 in Palfrey 1983, since the support of heavy-tailed distributions in Theorem 2.5.1 is the infinite positive semi-axis $\mathbf{R}_+$).

Theorem 2.5.1 *Suppose that the stand-alone valuations* X_i, $i \in M$, *for goods in* M *are i.i.d. positive stable r.v.'s with Lévy densities (2.2) with the index of stability* $\alpha = 0.5$. *Then buyers unanimously strictly prefer* (ex ante) $\underline{\mathcal{B}}$ *(that is, n separate auctions) to any other bundling decision.*

The intuition behind the results given by Theorem 2.5.1 is a reversal of the intuition for the results in Palfrey (1983). In the case of extreme heavy-tailedness, consumers' valuations per good for bundles become less concentrated about the mean as the size of bundles increases (see the discussion at the beginning of the section and the results in Sects. 1.3 and 2.1.4). Buyers who are on the upper tail of the distributions for the goods are more likely to win separate auctions and the next highest bidder is likely to have relatively lower valuations than in the case of a bundled auction. Therefore, contrary to the case of bounded valuations (see the discussion preceding Theorem 5 in Palfrey 1983) the winner of the auction is likely to prefer separate provision of the products.

We note in passing that in this section's setting with private values and the assumptions that valuations for each good as well as each agent are independently

---

[53]Clearly, in the case of discretely distributed valuations X_i, $i \in M$, consumers unanimously prefer $\mathcal{B}_1$ to $\mathcal{B}_2$ *ex ante* if each of them prefers $\mathcal{B}_1$ to $\mathcal{B}_2$ for all but a finite number of realizations of their stand-alone valuations.

distributed (so that interdependent valuations and affiliated signals—see Krishna 2002; Milgrom and Weber 1982—are ruled out). English ascending auctions over bundles are weakly equivalent to Vickrey auctions. Therefore, the results given by Theorem 2.5.1 continue to hold for English auctions as well.

As shown by Palfrey (1983), in Vickrey auctions with independently priced goods and an arbitrary number of bidders, the total surplus (that is, the sum of the seller's profit and buyers' surplus) is always maximized in the case when the goods are provided in separate auctions. The results in Palfrey (1983) imply that, under nonnegative valuations for individual goods and additive valuations for bundles, the seller prefers a single bundled Vickrey auction to any other bundling decision, if there are two buyers. The two buyers, on the other hand, unanimously prefer separate provision of items. The results for the two-buyer setting in Palfrey (1983) hold regardless of valuation distributions and therefore also for heavy-tailed valuations.

In the two-buyer setting, our results on consumers' preferences under heavy-tailed valuations are in accordance with Palfrey's. For more than two buyers, however, our results on the buyers' preferences differ from Palfrey's, who showed that (in the case of more than two buyers with bounded valuations) the buyers *never* unanimously prefer separate auctioning of the goods. By contrast, Theorem 2.5.1 shows that with heavy-tailed valuations they *always* unanimously prefer separate auctioning.

Theorem 2 in Palfrey (1983) shows that the seller always prefers to sell the goods in a single bundle when there are two buyers. This result holds regardless of distributions and therefore also when valuations are heavy-tailed. With more than two buyers, it is an open question what the optimal strategy for the seller is when distributions are heavy-tailed.

2.5.4 Optimal Bundling with Heavy-Tailed Valuations and Profit-Maximizing Prices

We turn to the case in which the prices for goods on the market and their bundles are set by the monopolist. To simplify the presentation of the results and their arguments, we assume that the marginal costs c_i of goods in M are identical: $c_i = c$, $i \in M$; however, extensions are possible for the case of arbitrary c_i. Suppose that the seller can provide bundles B of goods in M for prices per good $p \in [0, p_{\max}]$, where $p_{\max}$ is the (regulatory) maximum price, with the convention that $p_{\max}$ can be infinite. For a bundle of goods $B \in 2^M$, denote by p_B the profit-maximizing price per good for the bundle, so that the seller's expected profit from producing and selling bundles of B's (at the price p_B per good) is

$$E(\pi_B) = nk(p_B - c)P(v(B) \geq kp_B),$$

where $k = \text{card}(B)$. We focus on the pure bundling case. The profit maximizing price per good in the bundle is

$$p_B = \arg \max_{p \in [0, p_{\max}]} (p - c) P(v(B) \geq kp) = \arg \max_{p \in [0, p_{\max}]} (p - c) P\left(\sum_{i=1}^{k} X_i \geq kp \right).$$

Such optimization problems become much more complex in the mixed bundling case due to additional constraints in maximization for the buyers who can choose among many sets of bundles. We assume that $c < p_{\max}$, so that all bundles of goods in M are offered for sale. Clearly, in the case $c_i = c$ for all $i \in M$, the values of p_B are the same for all bundles B that consist of the same number $\text{card}(B)$ of goods. That is, $p_B = p_{B'}$, if $\text{card}(B) = \text{card}(B')$. Denote by $\bar{p}$ the profit maximizing price per good in the case where all the goods in M are sold as a single bundle and by $\underline{p}$ the profit maximizing price of each good $i \in M$ under unbundled sales. That is, $\bar{p} = p_B$ with $B = M$, and $\underline{p} = p_B$ with $B = \{i\}$, $i \in M$.

Theorems 2.5.2 and 2.5.3 characterize the optimal bundling strategies for a multiproduct monopolist with an arbitrary degree of heavy-tailedness of valuations for goods in M. From Theorem 2.5.2 it follows that if consumers' reservation prices are moderately heavy-tailed, then the patterns in seller's optimal bundling strategies are the same as in the case of independently priced goods with log-concavely distributed (thin-tailed) valuations (see Bakos and Brynjolfsson 1999; Fang and Norman 2006, and the discussion at the beginning of this section).

Theorem 2.5.2 is formulated for symmetric moderately heavy-tailed valuations. The results in the theorem further continue to hold for moderately heavy-tailed valuations with asymmetric stable distributions $S_\alpha(\sigma, \beta, \mu)$ with $\alpha > 1$ discussed in Sect. 2.1.2.

Theorem 2.5.2 *Let $\mu \in \mathbf{R}$. Suppose that the stand-alone valuations X_i, $i \in M$, for goods in M satisfy $X_i = \mu + \eta_i$, where η_i are i.i.d. r.v.'s such that $\eta_i \sim \overline{CS}$. The risk-neutral seller strictly prefers $\overline{B}$ to any other bundling decision (that is, the goods are sold as a single bundle), if $\underline{p} < \mu$. The risk-neutral seller strictly prefers $\underline{B}$ to any other bundling decision (that is, the goods are sold separately), if $\bar{p} > \mu$.*

Theorem 2.5.3 shows that the patterns in the solutions to the seller's optimal bundling problem in Theorem 2.5.2 are reversed if consumers' valuations are extremely heavy-tailed. Similar to Theorem 2.5.1, Theorem 2.5.3 is formulated for extremely heavy-tailed valuations with densities (2.2) and the tail index and the index of stability $\zeta = \alpha = 0.5$. The results continue to hold for all other extremely heavy-tailed stable valuations with stable distributions $S_\alpha(\sigma, \beta, \mu)$ in Sect. 2.1.2 with the indices of stability $\alpha < 1$ (in particular, the results hold for extremely heavy-tailed distributions $S_\alpha(\sigma, 1, \mu)$ with $\alpha < 1$ and $\beta = 1$ that are concentrated on the semi-axis $[\mu, \infty)$). In addition, they continue to hold for symmetric extremely heavy-tailed valuations $X_i = \mu + \eta_i$, where η_i are i.i.d. r.v.'s from the class $\underline{CS}$.

Theorem 2.5.3 *Let $\mu \in \mathbf{R}$ and $p_{\max} < \infty$. Suppose that the stand-alone valuations X_i, $i \in M$, for goods in M are i.i.d. positive stable r.v.'s with Lévy densities (2.2) with the index of stability $\alpha = 0.5$. The risk-neutral seller strictly prefers $\underline{\mathcal{B}}$ to any other bundling decision (that is, the goods are sold separately), if $\overline{p} < \mu$. The risk-neutral seller strictly prefers $\overline{\mathcal{B}}$ to any other bundling decision (that is, the goods are sold as a single bundle), if $\underline{p} > \mu$.*

Remark 2.5.1 Analogues of Theorems 2.5.2 and 2.5.3 hold for expected utility comparisons for a risk-averse seller, as long as her risk-aversion is not too high. Specifically, since the preferences over bundling decisions in Theorems 2.5.2 and 2.5.3 are strict, they will also hold for a slightly risk-averse seller. For a severely risk-averse seller, however, the results in Theorems 2.5.2 and 2.5.3 may not hold (see also Theorem 3 in Ibragimov and Walden 2007, for a discussion of diversification decisions in the VaR versus expected utility framework).

Remark 2.5.2 From property (2.3) it follows that, in the case of Cauchy valuations with densities (2.1) and $\alpha = 1$, $P\left(\sum_{i=1}^{k} X_i \geq kp \right) = P(X_1 \geq p)$ for all $1 \leq k \leq m$, and, consequently, $p_B = \underline{p} = \overline{p}$ and $E(\Pi_B) = E(\Pi_{\underline{B}}) = E(\Pi_{\overline{B}})$ for all bundling decisions $\mathcal{B}$. Thus, in the case of heavy-tailed valuations with the tail index (index of stability) $\alpha = 1$ as in the case of Cauchy valuations, the seller is indifferent among all bundling decisions in Theorems 2.5.2 and 2.5.3.

The condition $p_{\max} < \infty$ in Theorem 2.5.3 is necessary since otherwise the monopolist would set an infinite price for each bundle of goods under extremely heavy-tailed distributions of consumers' valuations considered in the theorem.

Similar to the argument based on variance in Bakos and Brynjolfsson (1999), the underlying intuition for Theorem 2.5.2 is that for moderately heavy-tailed distributions of reservation prices and the marginal costs of goods on the right of the mean valuation, bundling decreases profits since it reduces concentration (peakedness) of the valuation per good and thereby decreases the fraction of buyers with valuations for bundles greater than their total marginal costs (this is implied by the results in Sects. 2.1.3 and 2.1.4). For the identical marginal costs of goods less than the mean valuation, bundling is likely to increase profits.

On the other hand, similar to Vickrey auctions in Sect. 2.5.3, the results in Theorem 2.5.3 are driven by the fact that, in the case of extremely heavy-tailed reservation prices, concentration and peakedness of the valuations per good in bundles decreases with their size (see the results and discussion in Sects. 2.1.3 and 2.1.4 and also the discussion in the previous section). Therefore, bundling of goods in the case of extremely heavy-tailed valuations and marginal costs of goods higher than the mean reservation price increases the fraction of buyers with reservation prices for bundles greater than their total marginal costs and thereby leads to an increase in the monopolist's profit. This effect is reversed in the case of the identical marginal costs on the left of the mean valuation.

2.6   Economic Theory: Models of Firm Growth[54]

The goal of this section is to demonstrate that robustness of statistical procedures to heavy-tailedness can have important effects on the properties of firm growth models. Focusing on the model of demand-driven innovation and spatial competition over time in Jovanovic and Rob (1987), we show that the firms' growth patterns depend crucially on the degree of heavy-tailedness of consumers' signals and on the choice of estimators employed by the firms to make inferences about their markets. If consumers' signals in the model are extremely heavy-tailed and the firms use the sample means of the signals as product designs, then the firms' output levels exhibit anti-persistence and smaller firms have an advantage over their larger counterparts. These properties are reversals of those that hold under moderately heavy-tailed signals or in the case when the firms switch to more robust estimators of the ideal product, such as sample medians, in the presence of extreme heavy-tailedness.

2.6.1   Output Persistence and Demand-Driven Innovation and Spatial Competition Over Time

Since the seminal work of Nelson and Plosser (1982), many studies in economics have focused on models that could account for positive persistence in levels of output, among other "stylized facts" on output dynamics. Most of the models proposed in this stream of literature focus primarily on technology shocks as the driving force of economic fluctuations and usually rely on capital accumulation, intertemporal substitution, capital irreversibility or different types of capital adjustments costs or lags as sources of shock propagation to generate persistence.

Jovanovic and Rob (1987), hereafter JR, develop a model of demand-driven innovation and spatial competition over time in which the source of output persistence is, in contrast, private information alone. The model is based on the idea that larger firms get better information about their markets. The firms choose their products and then make output decisions based on how successful their product design is (in terms of the closeness to the ideal product). In the model, output decision has two effects. One is to maximize contemporaneous profits. The other is that output generates signals and thus information about the next period's ideal product. The greater is the output the more signals are likely to be received regarding the next period's ideal product and more information about the firm's market is likely to be collected.

---

[54]This section has drawn upon material from the article Ibragimov (2014) "On the robustness of location estimators in models of firm growth under heavy-tailedness," in press in the *Journal of Econometrics*, reproduced with permission from Elsevier, Copyright Elsevier (2014).

JR further focus on the analysis of the properties of the model in the case where the distribution of consumers' signals is log-concave which implies that the tails of signals' distributions decline at least exponentially fast and, thus, the distributions are thin-tailed, see Sect. 2.1.2. From the results in JR it follows that if the signals are log-concavely distributed and the firms use the sample mean of consumers' signals to estimate the ideal product (the center of the signals' distribution) and choose it as the product design, the model implies positive persistence in output levels. Furthermore, in such a setting, large firms always have an advantage over their smaller counterparts. More precisely, according to JR, under the above assumptions, the model has the following properties: the probability of rank reversals in adjacent periods (that is, the probability of the smaller of two firms becoming the larger one next period) is always less than one half; this probability diminishes as the current size-difference increases; and the distribution of future size is stochastically increasing as a function of current size. The intuition for the results is that the larger is a firm's size, the greater is the amount of information the firm gets. The larger firms that learn more are thus more likely to come up with a successful product.

One should note here that the analysis of (arbitrary) log-concavely distributed signals in JR implicitly makes the assumption that the firms choose the sample mean of consumers' signals as the product design. This is because the optimal product design in the setting employed in JR is the posterior median of the ideal product θ given a sample of signals S (see Eq. (6) in JR and their discussion at the beginning of Sect. 2.5). Although the posterior median coincides with the sample mean in the case of normal signals and diffuse priors, it is not the case in general, see, e.g., Sects. 4.2.1, 4.2.3, and 4.3.1 in Berger (1985), Sects. 3.1 and 3.2 in Box and Tiao (1973) and Chap. 2 and Sect. 4.2 in Carlin and Louis (2000) (under the normality assumption for the sample of observed signals S and a diffuse prior for θ, the sample mean of signals in S is the posterior median, mode and mean of θ).

The fact that a number of economic and financial time series have the tail exponents of approximately equal to or (slightly or even substantially) less than one discussed in Sect. 1.2 is important in the context of the results in section. As we demonstrate, the value of the tail index $\zeta = 1$ (that is, existence of the first moment) is exactly the critical boundary between robustness of implications of the model of demand-driven innovation and spatial competition over time to heavy-tailedness assumptions and their reversals. According to the results obtained in this section, the implications of the model are robust to heavy-tailedness assumptions with tail indices $\zeta > 1$ (Theorem 2.6.1). But its conclusions are reversed for extremely heavy-tailed distributions with $\zeta < 1$ and infinite first moments (Theorem 2.6.2).

We prove *inter alia* that if consumers' signals are independent and extremely heavy-tailed and the firms choose the sample mean of the signals as the product design then relatively large firms are not likely to stay larger and the model thus implies anti-persistence in output levels. In this case, a surprising pattern of oscillations in firm sizes emerges, with smaller firms being likely to become larger ones next period, and vice versa. Moreover, it is likely that very small firms will become very large next period, and the size of very large firms will shrink to very small.

More precisely, under the above assumptions, the probability of rank reversals in adjacent periods (that is, the probability of the smaller of the two firms becoming the larger one next period) is always greater than one half; this probability increases as the current size-difference increases; and the distribution of future size is stochastically decreasing as a function of current size.

Essentially, in the case of extremely heavy-tailed signals, smaller firms, in fact, have an advantage over their larger counterparts if the sample mean is employed as the product design. The driving force for this conclusion is that in the presence of extremely heavy-tailed shocks, the sample mean of signals is not informative about the ideal product (population center) θ since the sample of signals is very likely to contain extreme outliers. Sensitivity of the sample mean to the presence of extreme outliers also implies, according to our results, that if consumers' signals are extremely heavy-tailed, then it is optimal for the firms to switch to employing more robust estimators of the next period's product such as the sample median.

The assumption that the sample mean of signals is employed to approximate the ideal product (estimate the population center) and is chosen as the product design in the case of extremely heavy-tailed signals is appropriate in the setting where the firms do not realize that they are in the presence of extreme heavy-tailedness and utilize the same inference methods as in the case of distributions with thin tails. The firms might not be able to make inferences about heavy-tailedness of consumers' signals on their markets because of time or data availability constraints. The presence of heavy-tailedness and extreme signals, together with constraints on making inferences about it, is likely to be the case for industries with very uncertain consumer perception of new products or constantly changing environments and new industries in which business decisions on the basis of former experience are impossible and the risk facing the firms is higher than in other sectors. Many high-tech industries, together with the Net economy, exhibit the above patterns (see the discussion at the end of Sect. 2.6.4).

As follows from the results in Theorem 2.6.3 in the section, if the firms know that they are in the presence of extreme heavy-tailedness and employ robust inference methods, namely, use the sample median instead of the sample mean as the product design, then the counterintuitive conclusions discussed above disappear. According to Theorem 2.6.3, if the sample median is employed as the product design, then larger firms have an advantage over their smaller counterparts in the case of *arbitrary symmetric* consumers' signals. That is, in any such setting, the implications of the model of demand-driven innovation and spatial competition over time for the sample mean and log-concavely distributed signals in JR continue to hold.

The results obtained in this section highlight, therefore, the necessity of making inferences about the presence or absence of heavy-tailedness and extreme outliers before making business decisions, if possible, and of employing robust estimation methods, such as the use of the sample median instead of the sample mean in the presence of heavy-tailed signals. According to the results, having more information is always advantageous if robust inference methods are employed; this advantage, however, can be completely lost and even become a disadvantage if the decisions are made using non-robust estimators in the presence of extreme heavy-tailedness.

2.6.2 Growth Theory for Firms Investing into Information About Their Markets

In this and in the next section, we review the setup of the model of demand-driven innovation and spatial competition over time developed by JR and its properties under log-concavity of signals' distributions.

Consider a market for a differentiated commodity. Let $\hat{\theta} \in \mathbf{R}_+$ be a location variable which differentiates the firm's product, let $\theta \in \mathbf{R}_+$ be an "ideal" product, and let $\rho(x, \theta) = |x - \theta|$, $x \in \mathbf{R}_+$, denote the absolute loss function.[55] A consumer of type $u \in \mathbf{R}_+$ has the utility function $u - \rho(\hat{\theta}, \theta) - p_{\hat{\theta}}$, if she purchases one unit of good produced by the firm, and 0, if not, where $p_{\hat{\theta}}$ is the price the consumer pays for the good. Consumers are assumed to be perfectly informed about all price-quality combinations offered by various sellers and the firm is assumed to be a price taker. In what follows, we suppose that the price p of the ideal product θ is unity in terms of some "outside good": $p = 1$.

Under the above assumptions, a necessary condition for an equilibrium is that $\rho(\hat{\theta}, \theta) + p_{\hat{\theta}} = 1$ for all $\hat{\theta} \in \mathbf{R}$.

Simplifying the setting of the model considered in JR, we suppose that each period the firm makes two decisions. First, it chooses the product design $\hat{\theta}$, and does so before knowing what θ prevails for that period. The commitment to a particular $\hat{\theta}$ is costless but irreversible until next period. Having committed to $\hat{\theta}$, the firm then learns θ. Being of measure zero, the firm will be a price taker and its price is

$$p_{\hat{\theta}} = 1 - \rho(\hat{\theta}, \theta). \tag{2.37}$$

The firm then chooses the level of output y, with $C(y)$ denoting the corresponding convex and twice differentiable cost function.

Each period, the firm observes a sample $\mathcal{S}$ of signals $s_i = \theta + \epsilon_i$, $i = 1, \ldots, N$, about the next period's ideal product $\theta \in \mathbf{R}$, where ϵ_i, $i = 1, \ldots, N$, are i.i.d. unimodal shocks with mode 0 and N is a (random) sample size. The size N of the sample $\mathcal{S}$ of signals about the next period's ideal product observed by the firm follows a distribution $\pi(n; y)$ conditionally on y : $\pi(n; y) = P(N = n|y)$, $n = 0, 1, 2, \ldots$ The function $\pi(n; y)$ is assumed to be increasing in y for all n, so that N is stochastically increasing in y and larger firms are likely to get more signals each period and to learn more about the next period's ideal product.

---

[55]From the proof of the results in this section it follows that they continue to hold in the case of arbitrary loss functions $\rho(x, y) = \psi(|x - y|)$, where ψ is nonnegative and increasing on $\mathbf{R}_+$.

Below, we denote by S_t, $\hat{\theta}_t$, θ_t and y_t the values of the variables in period t. In the model, the sequence of events is as follows: in period t, first S_t is observed, next $\hat{\theta}_t$ is chosen; then θ_t is observed and y_t is chosen; the period then ends.

Throughout the section, for $v = (v_1, v_2, \ldots, v_n) \in \mathbf{R}^n$, we denote by $\bar{v}_n = g_1(v_1, \ldots, v_n) = n^{-1} \sum_{i=1}^{n} v_i$ the sample mean of v_i's. In the case when n is odd, $n = 2k-1$, we further denote by $\tilde{v}_n$ the sample median (that is, the kth order statistic) of $v_1, \ldots, v_n : \tilde{v}_n = g_2(v_1, \ldots, v_n) = v_{(k)} = median(v_1, v_2, \ldots, v_n)$ (here and in what follows, $v_{(1)} \leq v_{(2)} \leq \ldots \leq v_{(n)}$ stand for components of x in nondecreasing order).

Let $g(v) = g(v_1, v_2, \ldots, v_n)$ be an estimator based on a sample of observations $v = (v_1, v_2, \ldots, v_n) \in \mathbf{R}^n$ that satisfies the translation equivariance condition: $g(v_1 + a, v_2 + a, \ldots, v_n + a) = g(v_1, v_2, \ldots, v_n) + a$ for all $a \in \mathbf{R}$ (see Bickel and Lehmann 1975a,b, Chap. 4 in Rousseeuw and Leroy 1987, and Sects. 2.3 and 2.4 in Jurečková and Sen 1996). Evidently, this condition holds for the sample mean $g_1(v_1, \ldots, v_n) = \bar{v}_n$, $n \geq 1$, and the sample median $g_2(v_1, \ldots, v_n) = \tilde{v}_n$, $n = 2k - 1$, $k = 1, 2, \ldots$

Let $F(x; n) = P(|g(\epsilon_1, \epsilon_2, \ldots, \epsilon_n)| \leq x)$, $x \geq 0$, $n = 1, 2, \ldots$, denote the cdf of $|g(\epsilon_1, \epsilon_2, \ldots, \epsilon_n)|$, $n = 1, 2, \ldots$, on $\mathbf{R}_+$, so that $F(x; n) = P(|\bar{\epsilon}_n| \leq x)$, $n = 1, 2, \ldots$, for the sample mean $g_1(\epsilon_1, \ldots, \epsilon_n) = \bar{\epsilon}_n$, and $F(x; n) = P(|\tilde{\epsilon}_n| \leq x)$, $n = 2k - 1$, $k = 1, 2, \ldots$, for the sample median $g_2(\epsilon_1, \ldots, \epsilon_n) = \tilde{\epsilon}_n$.

Suppose that, for $N > 0$, the firm chooses the estimator $\hat{\theta} = \hat{\theta}(S) = g(s_1, \ldots, s_N)$ of θ as the product design. The loss associated with this choice of $\hat{\theta}$ for $N > 0$ is $\rho(\hat{\theta}(S), \theta) = |g(\epsilon_1, \epsilon_2, \ldots, \epsilon_N)|$. In the case when $N = 0$ belongs to the support of N, so that $\pi(0; y) \neq 0$, it is usually assumed that $\rho(\hat{\theta}(S), \theta) = \infty$ for $N = 0$. The cdf of $\rho(\hat{\theta}(S), \theta)$ (on $\mathbf{R}_+$) conditional on y is

$$\xi(x; y) = P(\rho(\hat{\theta}(S), \theta) \leq x|y) = \sum_{n=0}^{\infty} F(x; n)\pi(n; y), \qquad (2.38)$$

$x \geq 0$ (with $F(x; 0) = 0$ if $N = 0$ belongs to the support of N under the above convention).

Each period, the firm wishes to maximize expected profits discounted by the factor β. The dynamic programming formulation of the firm's problem of choosing y following a realization $\rho(\hat{\theta}, \theta) = x$, is the (recursive) Bellman equation

$$V(x) = \max_y \left\{ y(1 - x) - C(y) + \beta \int V(\tilde{x}) d\xi(\tilde{x}; y) \right\}$$

(see also JR for details).

Let $G(y) = \beta \int V(\tilde{x}) d\xi(\tilde{x}; y)$. The first-order and second-order conditions for an interior maximum in y are

$$p_{\hat{\theta}} - C'(y) + G'(y) = 0, \qquad (2.39)$$

$$G''(y) < C''(y). \qquad (2.40)$$

We assume that, for any continuous $f : \mathbf{R} \to \mathbf{R}$, the expression $\int f(\tilde{x}) d\xi(\tilde{x}; \lambda)$ is differentiable in λ. Under this assumption, one can implicitly differentiate first-order condition (2.39) (see JR).

Evidently, condition (2.40) holds if the function $G(y)$ is strictly concave: $G'' < 0$. However, $G'' > 0$ (strictly convex G) is also consistent with maxima being interior.[56]

2.6.3 Log-Concave Signals and Demand-Driven Innovation and Spatial Competition Over Time

Throughout the section, the distribution $\pi(n; y) = P(N = n|y)$ of N conditional on y will be assumed to be one of the following: a Poisson distribution with the mean $\mu y : \pi_0(n; y) = \frac{(\mu y)^n}{n!} exp(-\mu y), n = 0, 1, \ldots$ (with the convention that $\rho(\hat{\theta}, \theta) = \infty$ for $N = 0$); a shifted Poisson distribution $\pi_1(n; y) = \frac{(\mu y)^{n-1}}{(n-1)!} exp(-\mu y), n = 1, 2, \ldots$; or a Poisson-type distribution concentrated on odd numbers $\pi_2(n; y) = \frac{(\mu y)^{k-1}}{(k-1)!} exp(-\mu y)$ for $n = 2k - 1, k = 1, 2, \ldots, \pi_2(n; y) = 0$ for $n = 2k$, $k = 0, 1, 2, \ldots$ (note that there is no ambiguity concerning the value of $\rho(\hat{\theta}, \theta)$ in the case $N = 0$ for distributions π_1 and π_2). The supports of the distributions $\pi_j, j = 0, 1, 2$, are, respectively, $M_0 = \{0, 1, 2, \ldots, \}$, $M_1 = \{1, 2, 3, \ldots, \}$ and $M_2 = \{1, 3, 5, \ldots, \}$.

In this section, we consider the conclusions of the model of demand-driven innovation and spatial competition over time in the case where the firm employs the sample means $\bar{s}_N$ as in JR (see the discussion in Sect. 2.6.1) or sample medians $\tilde{s}_N$ of consumers' signals $s_1, \ldots, s_N$ as product designs: $\hat{\theta} = \hat{\theta}(\mathcal{S}) = g_1(s_1, \ldots, s_N) = \bar{s}_N$ or $\hat{\theta} = \hat{\theta}(\mathcal{S}) = g_2(s_1, \ldots, s_N) = \tilde{s}_N$ for $N = 2k - 1, k = 1, 2, \ldots$

JR obtained the following Proposition 2.6.1.[57] In the proposition and its analogues for heavy-tailed signals obtained below (Theorems 2.6.1 and 2.6.2), $y_t^{(1)}$ and $y_t^{(2)}$ are sizes of two firms at period t; $y_{t+1}^{(1)}$ and $y_{t+1}^{(2)}$ stand for their sizes next period.

Proposition 2.6.1 (JR) *Suppose that, conditionally on y, N has one of the distributions $\pi_j(n; y), j = 0, 1, 2$. Let the shocks $\epsilon_1, \epsilon_2, \ldots$ be i.i.d. r.v.'s such that $\epsilon_i \sim \mathcal{LC}$, $i = 1, 2, \ldots$ If the optimal level y_t of output satisfies the first- and second-order conditions for an interior maximum (2.39) and (2.40) and the firm chooses the*

---

[56]By Proposition 4 in JR, in the model of demand-driven innovation and spatial competition over time involving the choice of informational gathering effort z in addition to the choice of output y, larger firms always invest more in information if the function G is convex ($G'' > 0$). Thus, under this condition, investment z into gathering information in JR is secondary with respect to persistence results comparing to y. One should note that, according to empirical studies, there is a positive relationship between R&D expenditures and firm size, that suggests that G is indeed convex (see Kamien and Schwartz 1982, and the discussion following Proposition 4 in JR).

[57]In JR, the proposition is formulated for the Poisson distribution π_0. The argument for the distributions $\pi_j, j = 1, 2$, is completely similar to that case.

sample mean $\hat{\theta} = g_1(s_1, \ldots, s_N) = \bar{s}_N$ as the product design for $N > 0$, then the following conclusions (a)–(c) hold.

(a) The probability of rank reversals in adjacent periods $P(y_{t+1}^{(1)} > y_{t+1}^{(2)} | y_t^{(2)} > y_t^{(1)})$ is always less than $1/2$.
(b) This probability diminishes as the current size-difference $y_t^{(2)} - y_t^{(1)}$ increases (holding constant the size of one of the firms).
(c) The distribution of future size is stochastically increasing as a function of current size y_t, that is, $P(y_{t+1} > y | y_t)$ is increasing in y_t for all $y \geq 0$.

Lemma 2.6.1 in JR and its proof imply the following *sufficient* conditions for concavity of the function $G(y)$; under the assumptions of the lemma, therefore, the second-order condition (2.40) for an interior maximum with respect to y is satisfied.

Lemma 2.6.1 (JR) *Suppose that, conditionally on y, N has one of the distributions $\pi_j(n; y)$, $j = 0, 1, 2$. The function $G(y)$ is strictly concave in y if the sequence $\{F(x; n+1) - F(x; n)\}_{n=0}^{\infty}$ is strictly decreasing in n for all $x > 0$.*

As noted in JR, the conditions of Lemma 2.6.1 are satisfied for normal r.v.'s $\epsilon_i \sim \mathcal{N}(0, \sigma^2)$, $i = 1, 2 \ldots$ and the sample means $\bar{s}_N$ employed as product designs.

2.6.4   Main Results: Robustness to Heavy-Tailedness Assumptions

In this section, we present the main results of the section on the robustness of the model of demand-driven innovation and spatial competition over time to heavy-tailedness assumptions and the choice of location estimators as product designs. The results are formulated for the case of symmetric heavy-tailed signals from the classes $\overline{\mathcal{CSLC}}$ and $\mathcal{CSLC}$. They also continue to hold for signals with heavy-tailed asymmetric stable distributions $S_\alpha(\sigma, \beta, \mu)$ discussed in Sect. 2.1.2.

The following theorem provides a generalization of Proposition 2.6.1 that shows that the results obtained by JR continue to hold in the case of moderately heavy-tailed signals.

Theorem 2.6.1 *Suppose that, conditionally on y, N has one of the distributions $\pi_j(n; y)$, $j = 0, 1, 2$. Let the shocks $\epsilon_1, \epsilon_2, \ldots$ be i.i.d. r.v.'s such that $\epsilon_i \sim \overline{\mathcal{CSLC}}$, $i = 1, 2, \ldots$ Then conclusions (a), (b), and (c) in Proposition 2.6.1 hold.*

Lemma 2.6.2 shows that strict concavity of the function $G(y)$ in Lemma 2.6.1 and, consequently, the second-order condition (2.40) are satisfied for shocks $\epsilon_1, \epsilon_2, \ldots$ with moderately heavy-tailed symmetric stable distributions and the sample mean of signals employed as the product design.

Lemma 2.6.2 *Suppose that the firm chooses the sample mean $\hat{\theta} = g_1(s_1, \ldots, s_N) = \bar{s}_N$ as the product design for $N > 0$ and the shocks $\epsilon_1, \epsilon_2, \ldots$ are i.i.d. r.v.'s such* *that $\epsilon_i \sim S_\alpha(\sigma)$, $i = 1, 2, \ldots$, for some $\sigma > 0$, and $\alpha \in (1, 2]$. Then the sequence $\{F(x; n+1) - F(x; n)\}_{n=0}^\infty$ in Lemma 2.6.1 is strictly decreasing in n for all $x > 0$. Thus, the function $G(y)$ is strictly concave in y if, conditionally on y, N has one of the distributions $\pi_j(n; y)$, $j = 0, 1, 2$.*

As the following theorem shows, the conclusions of Proposition 2.6.1 and Theorem 2.6.1 are reversed in the case of shocks $\epsilon_1, \epsilon_2, \ldots$ with extremely heavy tails.

Theorem 2.6.2 *Suppose that, conditionally on y, N has one of the distributions $\pi_j(n; y)$, $j = 1, 2$. Let the shocks $\epsilon_1, \epsilon_2, \ldots$ be i.i.d. r.v.'s such that $\epsilon_i \sim \underline{CS}$, $i = 1, 2, \ldots$ If the optimal level y_t of output satisfies the first- and second-order conditions for an interior maximum (2.39) and (2.40) and the firm chooses the sample mean $\hat{\theta} = g_1(s_1, \ldots, s_N) = \bar{s}_N$ as the product design for $N > 0$, then the following conclusions (a')–(c') hold.*

(a') *The probability of rank reversals in adjacent periods $P(y_{t+1}^{(1)} > y_{t+1}^{(2)} | y_t^{(2)} > y_t^{(1)})$ is always greater than 1/2.*

(b') *This probability increases as the current size-difference $y_t^{(2)} - y_t^{(1)}$ increases (holding constant the size of one of the firms).*

(c') *The distribution of future size is stochastically decreasing as a function of current size y_t, that is, $P(y_{t+1} > y | y_t)$ is decreasing in y_t for all $y \geq 0$.*

Remark 2.6.1 From the proof of Theorem 2.6.2 it follows that, under its assumptions, $G' \leq 0$. It is not difficult to see that this implies that, in the setting of JR's model with the choice of investment z into information gathering in addition to the choice of quantity y, the optimal level of z is zero if the investment cost $K(z)$ is increasing and the first- and second-order conditions for an interior maximum are satisfied.

According to our results, there is no informational advantage in the presence of extremely heavy-tailed signals if the sample mean is used as the product design. As the following theorem shows, having more signals is, however, always advantageous if a more robust estimator of θ, namely, the sample median, is used as the product design instead of the sample mean. Moreover, this conclusion holds for arbitrary symmetric consumers' signals.

Theorem 2.6.3 *Suppose that, conditionally on y, N has the Poisson-type distribution $\pi_2(n; y)$ and $\epsilon_1, \epsilon_2, \ldots$ are i.i.d. r.v.'s with a symmetric density $f(x)$. If the optimal level y_t of output satisfies (2.39) and (2.40) and the firm chooses the sample median $\hat{\theta} = g_2(s_1, \ldots, s_N) = \tilde{s}_N$ as the product design for $N = 2k-1$, $k = 1, 2, \ldots$, then conclusions (a)–(c) of Proposition 2.6.1 hold.*

The results in this section provide new insights concerning firm size and growth patterns in different industries. According to the results in Audretsch (1991), survival rates for incumbents are higher than for entrants in markets for "nontechnical" products, where advantages of experience and learning by doing are high. These conclusions are similar to advantages of large firms over small ones in the model of demand-driven model of innovation and spatial competition over time with moderately heavy-tailed consumers' signals in Theorem 2.6.1. On the other hand, Agarwal and Gort (1996) have observed that entrants in markets for high-technology products tend to have higher survival rates than incumbents. These results for "technical" products are consistent with findings in Audretsch (1991) who showed that new firm survival rates tend to be higher in sectors with high innovative activity by small firms, which are more likely to be recent entrants. Agarwal and Gort (1996) note that their results are consistent with entry in high technology industries being accompanied by breakthroughs in knowledge or innovations by inventors and firms initially outside the market. Such breakthroughs in knowledge are extreme events that yield superior knowledge to entrants and give them an advantage over incumbents, similar to an advantage of small firms over their large counterparts in the model of demand-driven model of innovation and spatial competition over time with extremely heavy-tailed signals in Theorem 2.6.2. The oscillation patterns in the firm sizes predicted by the results for extremely heavy-tailed signals in Theorem 2.6.2 might be further illustrated by the rapid rise of Internet businesses during the late 1990s and their sudden fall following an extreme event, the fall of NASDAQ by 10 % in April, 2000.

2.7 Summary and Conclusions

The results presented in this chapter demonstrate that heavy-tailedness is of key importance for robustness of many models in economics, finance, risk management, insurance, econometrics, and statistics. Importantly, its presence may either reinforce or reverse the properties and implications of key models in these fields, depending on the degree of heavy-tailedness. This includes diversification analysis, the properties of (re-)insurance markets for catastrophe risks, the models of risk-sharing, optimal bundling strategies for a multiproduct monopolist, firm growth theory models and the properties of important econometric and statistical inference procedures, among others. The models considered in this chapter have the value of the tail index $\zeta = 1$ as the critical boundary between robustness and reversals of their properties under heavy-tailedness. In other words, the models are robust to moderate heavy-tailedness with tail indices ζ greater than one and finite first moments. Their properties and conclusions are reversed in presence of extreme heavy-tailedness with tail indices ζ smaller than one and infinite first moments.

The results in the chapter further emphasize importance of econometrically and statistically justified inference, e.g., using correct standard errors and confidence intervals, for the tail indices ζ and the degree of heavy-tailedness. The analysis of the degree of uncertainty in estimation of tail indices ζ, their standard errors and confidence intervals is of key importance in deciding whether the standard economic and financial models and classical econometric and statistical methods are applicable. This is especially so in markets where there may be a potential for non-robustness due to possible presence of heavy tails with $\zeta < 1$, and thus the tests of the hypothesis $\zeta = 1$ vs. $\zeta < 1$ and $\zeta > 1$ are of key interest. Naturally, $\zeta = 2$ is another tail index value important in inference on the degree of heavy-tailedness as it is the critical boundary between finite and infinite second moments and well-defined vs. undefined variances: The tests of the hypothesis $\zeta = 2$ vs. $\zeta < 2$ and $\zeta > 2$ using correct standard errors and confidence intervals on ζ are of crucial importance for applicability of classical econometric and statistical inference methods, including the regression and least squares analysis and (auto-)correlation based and related approaches (see the discussion in Sects. 1.2 and 2.1.4). In the case of applications of the standard autocorrelation based methods in the analysis of financial returns and foreign exchange rates that are often modelled using GARCH and related processes, the property whether $\zeta < 4$ becomes of crucial importance (see Cont 2001; Davis and Mikosch 1998; Mikosch and Stărică 2000). Further, testing whether $2 < \zeta < 4$ for a particular (e.g., emerging or developing) financial or foreign exchange market is of key importance in the analysis whether its heavy-tailedness properties are similar to those implied by empirical results that are well-established in the case of developed markets (see the review and discussion in Sect. 1.2 and its references).

The goal of the next chapter is to review econometrically and statistically justified approaches to inference on tail indices and the degree of heavy-tailedness based on widely used Hill's estimates and log-log rank-size regressions. The chapter provides several empirical applications for estimating tail indices in foreign exchange rate markets across countries. In addition, it further discusses recently developed methods for robust inference in the presence of heterogeneity, dependence, and heavy-tailedness of largely unknown form.

Of course, inferences about tail indices will typically not be conclusive. Instead, they will imply a range of possible indices (e.g., confidence intervals), in contrast to the models presented in this chapter, which assume a known tail index. An extension would be to consider risks that are Pareto law distributed, but with a particular distribution for their possible tail index. Given such a distribution of the tail index, with support on $[\underline{\alpha}, \infty)$ for some $\underline{\alpha} > 0$ and such that $P(\alpha \leq \underline{\alpha} + \epsilon) > 0$ for all $\epsilon > 0$, it is straightforward to show that the unconditional (on α) risk distribution is of Pareto type with the tail index $\underline{\alpha}$. The results in this section could therefore be extended to such situations with unknown tail indices.

Chapter 3
Inference and Empirical Examples

3.1 Inference on Heavy Tails[1]

Several approaches to the inference about the tail index ζ of heavy-tailed distributions are available in the literature (see, among others, the reviews in Beirlant et al. 2004; Embrechts et al. 1997). The two most commonly used ones are Hill's estimator and the OLS approach using the log-log rank-size regression.

Let $r_1, r_2, \ldots, r_N$ be a sample from a population satisfying power law (1.3). Further, let, for $n < N$,

$$|r|_{(1)} \geq |r|_{(2)} \geq \cdots \geq |r|_{(n)} \geq |r|_{(n+1)} \tag{3.1}$$

be decreasingly ordered largest absolute values of observations in the sample. Hill's estimator $\hat{\zeta}_{\text{Hill}}$ of the tail index ζ is given by

$$\hat{\zeta}_{\text{Hill}} = \frac{n}{\sum_{t=1}^{n} \left(\log |r|_{(t)} - \log |r|_{(n+1)} \right)}. \tag{3.2}$$

The standard error of the estimator is s.e.$_{\text{Hill}} = \frac{1}{\sqrt{n}} \hat{\zeta}_{\text{Hill}}$. The corresponding 95 %-confidence interval for the true tail index ζ is thus given by

$$\left(\hat{\zeta}_{\text{Hill}} - \frac{1.96}{\sqrt{n}} \hat{\zeta}_{\text{Hill}}, \hat{\zeta}_{\text{Hill}} + \frac{1.96}{\sqrt{n}} \hat{\zeta}_{\text{Hill}} \right). \tag{3.3}$$

---

[1]This section draws upon material from Gabaix and Ibragimov (2011) "Rank−1/2: A simple way to improve the OLS estimation of tail exponents," *Journal of Business and Economic Statistics*, Vol. 29, No. 1, 24–39.

© Springer International Publishing Switzerland 2015
M. Ibragimov et al., *Heavy-Tailed Distributions and Robustness in Economics and Finance*, Lecture Notes in Statistics 214, DOI 10.1007/978-3-319-16877-7_3

It was reported in a number of studies that inference on the tail index using Hill's estimator suffers from several problems, including sensitivity to dependence and small sample sizes (see, among others, Embrechts et al. 1997, Chap. 6). Motivated by these problems, several studies have focused on alternative approaches to the tail index estimation. For instance, Huisman et al. (2001) propose a weighted analogue of Hill's estimator that was reported to correct its small sample bias for sample sizes less than 1,000. Embrechts et al. (1997), among others, advocate sophisticated nonlinear procedures for tail index estimation.

Despite the availability of more sophisticated methods, a popular way to estimate the tail index ζ is still to run the following OLS log-log rank-size regression with $\gamma = 0$:

$$\log (t - \gamma) = a - b \log |r|_{(t)}, \tag{3.4}$$

$t = 1, \ldots, n$, or (calling t the rank of an observation, and $|r|_{(t)}$ its size): $\log (\text{Rank} - \gamma) = a - b \log (\text{Size})$, and take b as an estimate of the tail index (here and throughout the section, $\log(\cdot)$ stands for the natural logarithm). Similar log-log rank-size regressions applied to positive and negative observations r_t in the sample are employed to estimate the tail indices ζ_1 and ζ_2 in (1.1) and (1.2). The reason for the popularity of the OLS approaches to tail index estimation is arguably the simplicity and robustness of these methods. In various frameworks, the log-log rank-size regressions of form (3.4) in the case $\gamma = 0$ and closely related procedures were employed, in particular, in Levy (2003), Levy and Levy (2003), Helpman et al. (2004), and many other works (see also the review and references in Gabaix and Ibragimov 2011).

Unfortunately, tail index estimation procedures based on OLS log-log rank-size regressions (3.4) with $\gamma = 0$ are strongly biased in small samples. The recent study by Gabaix and Ibragimov (2011) provides a simple practical remedy for this bias, and argues that, if one wants to use an OLS regression, one should use the Rank $-1/2$, and run $\log (\text{Rank} - 1/2) = a - b \log (\text{Size})$, that is,

$$\log (t - 1/2) = a - b \log |r|_{(t)}, \tag{3.5}$$

$t = 1, \ldots, n$. In (3.5), one takes the OLS estimate $\hat{b}$ as the log-log rank-size regression estimate $\hat{\zeta}_{\text{RS}}$ of the tail index ζ. The shift of $1/2$ is optimal, and reduces the bias to a leading order. The standard error of the estimator $\hat{\zeta}_{\text{RS}}$ is s.e.$_{\text{RS}} = \sqrt{\frac{2}{n}} \hat{\zeta}_{\text{RS}}$ (the standard error is thus different from the OLS standard error).[2] The corresponding 95 % confidence interval for the true tail index ζ (denoted by

---

[2]Similar to the analysis in Gabaix and Ibragimov (2011), one can also obtain the results on the standard error on the estimate a of the constant term in the above log-log rank-size regression (evidently, a is an estimate of the logarithm of the scaling constant C in heavy-tailed population model (1.3)). Together with estimates and standard errors on the tail index ζ, these results can be used in calculating loss exceedance probabilities and in assessing commonly used risk measures

95 %; CI_{RS} in estimation results in this section) is

$$\left(\hat{\zeta}_{RS} - 1.96 \times \sqrt{\frac{2}{n}}\hat{\zeta}_{RS}, \hat{\zeta}_{RS} + 1.96 \times \sqrt{\frac{2}{n}}\hat{\zeta}_{RS}\right). \tag{3.6}$$

Numerical results in Gabaix and Ibragimov (2011) demonstrate the advantage of the proposed approach over the standard OLS estimation procedures (3.4) with $\gamma = 0$ and further show that it performs well under deviations from power laws and heavy-tailed dependent GARCH processes that are often used for modeling financial returns, exchange rates, and other important economic and financial variables in different markets. The modifications of the OLS log-log rank-size regressions with the optimal shift $\gamma = 1/2$ and the correct standard errors provided by Gabaix and Ibragimov (2011) were subsequently used in Bosker et al. (2007), Bosker et al. (2008), Gabaix and Landier (2008), Ioannides et al. (2008), Le Gallo and Chasco (2008), Zhang et al. (2009), di Giovanni et al. (2011), Acemoglu et al. (2012), Chollete et al. (2012), Hinloopen and van Marrewijk (2012), Toda (2012) and several other works. Due to inherent heterogeneity and dependence properties and data availability constraints, foreign exchange rates in emerging and developing markets provide natural areas for applications of robust inference methods. The next section provides the empirical analysis of heavy-tailedness in emerging country foreign exchange markets using the above log-log rank-size regressions with correct standard errors and optimal shifts in ranks developed in Gabaix and Ibragimov (2011).

3.2 Empirical Illustrations: Heavy-Tailedness in Emerging Foreign Exchange Markets[3]

Foreign exchange markets are arguably the world's largest markets, operating continuously, and bringing together a wide variety of buyers and sellers, within and across national borders. In recent years these markets have been characterized by turbulence and volatility, with extreme variations marking some exchange rates. As the literature on the determination of exchange rates points out, there are many processes capable of generating extreme exchange rate variations. These include economic crises, speculative attacks, bailouts, stabilization efforts, regime reforms and regulatory changes, among others. Recent theoretical literature contains useful models that explain extreme changes in financial returns, in terms of trading actions

---

such as the value at risk and expected shortfall relatively far in the tails of heavy-tailed distributions considered (see Sects. 1.2 and 2.1).

[3]This section is based on the article Ibragimov et al. (2013), which was published in the *Journal of Banking and Finance*, Volume 37, Issue 7, pp. 2546–2559, reproduced with permission from Elsevier, Copyright Elsevier (2013).

of large market participants (see, for instance, Gabaix et al. 2006), and in terms of government interventions in the case of foreign exchange markets.

Large fluctuations in exchange rates carry significant real consequences for international trade, foreign investment, asset prices, and a wide range of other economic and financial outcome variables. The on-going financial and economic crisis has raised the need for accurate estimates of probabilities associated with large changes in financial returns and exchange rates. Emerging and developing countries are generally held to be subject to more frequent and more pronounced external and internal shocks than their developed counterparts, and in that context it is ever more important to identify currencies that are relatively more prone to large fluctuations.

To our knowledge there are very few studies on heavy-tailedness properties in emerging and developing economies. Akgiray et al. (1988) focus on maximum likelihood estimation in parametric stable and Generalized Pareto power law families fitted to monthly observations for a number of Latin American exchange rates. The confidence intervals obtained using Hill's estimator suggest that the variance and even first moments of these time series may be infinite (see also Fofack and Nolan 2001, for maximum likelihood estimates for infinite variance stable distributions fitted to different exchange rates). Koedijk et al. (1992) estimate tail indices for Latin American exchange rates and find evidence for different tail behavior in exchange rate returns under different exchange rate regimes. The analysis in Akgiray et al. (1988) and Koedijk et al. (1992) is based on relatively small samples of monthly observations and therefore has wide confidence intervals. Quintos et al. (2001) develop tests for structural breaks in the tail index, and motivated by the Asian financial crisis, apply these tests to emerging Asian stock prices. Using extensions of tests in Quintos et al. (2001) to allow for multiple tail index breaks, Candelon and Straetmans (2006) focus on changes in the tail indices of six emerging Asian currencies (Indonesian Rupiah, Malaysian Ringgit (MYR), Thai Baht (THB), Philippine Peso, South Korean Won (KRW), and Pakistan Rupee) and five developed currencies (Japanese Yen (JPY), British Pound, Swiss Frank, Canadian Dollar (CAD), and German Mark) over the period from the beginning of 1994 to the middle of 2003.[4] The empirical results in Candelon and Straetmans (2006) point to statistically significant changes in the tail indices of exchange rates of most of the above emerging currencies over the 1997 Asian crisis period (with tail index drops corresponding to increases in the degrees of heavy-tailedness). In a number of cases the tail index breaks can be linked to changes in monetary and exchange rate policies. In contrast, statistically significant breakpoints are not observed in the tail indices of exchange rates of the developed countries. The estimates for (relatively small) samples of quarterly data on exchange rates in Asian, Latin American, and European economies in Pozo and Amuedo-Dorantes

---

[4]See also Payaslioğlu (2009), for applications of tests in Quintos et al. (2001) in the analysis of structural breaks in the tail index of the exchange rate in Turkey over periods with different foreign exchange regimes.

(2003) produce confidence intervals that indicate that the variances of the time series considered may be infinite.

Our principal goal in this section is the robust analysis of heavy-tailedness properties of exchange rates of emerging and developing countries, in comparison with developed countries. This comparative examination is motivated by the generally held view that the former set of countries are more subject to severe external and internal shocks, and therefore suffer greater potential for extreme changes in financial returns and exchange rates. We use recently proposed robust tail index estimation methods, based on log-log rank-size regressions with optimal shifts in ranks, and correct standard errors (see Sect. 3.1), applying them to large data sets on daily exchange rates for a number of countries. This is in contrast to earlier studies of exchange rates of emerging countries which have tended to use model-specific parametric maximum likelihood procedures or (semiparametric) Hill's estimators, with a number of contributions using relatively small data sets, with potentially non-robust conclusions.[5,6]

A further dimension to our analysis is the on-going economic crisis. We assess whether the crisis led to significant changes in the likelihood of large variations in exchange rates. We also draw conclusions on the applicability of standard economic and econometric models, including regression methods, and models explaining heavy tails in financial markets.

We find that the tail indices for exchange rates of emerging countries are indeed considerably smaller than those of developed countries. Our estimates imply that, in contrast to developed countries, the value of the tail index $\zeta = 2$ is not rejected at commonly used statistical significance levels for the exchange rates of several emerging countries (Sect. 3.2.2), implying that their variances may be infinite. Tail index values $\zeta = p \in (2.6, 2.8)$ appear to be at the dividing boundary between developed country exchange rates on the one hand, and emerging country exchange rates on the other: while the moments of order $p \in (2.6, 2.8)$ are finite for most of the developed country exchange rates, they may be (or are) infinite for most of the emerging country exchange rates.

With respect to the on-going financial and economic crisis, we find that while the heavy-tailedness properties of most exchange rates did not change significantly, a few foreign exchange markets did see structural changes. There was significant increase in the degree of heavy-tailedness of the Swiss franc (CHF) and pound

---

[5]Robustness of the tail index estimation approaches based on log-log rank-size regressions is illustrated by their favorable performance under deviations from power laws (1.1)–(1.3) in the form of slowly varying factors and dependent GARCH processes that are often used for modeling financial returns, exchange rates and other economics and financial time series (see the discussion in Sect. 3.1).

[6]For illustration, we compare the tail index estimates obtained using the log-log rank-size regression approach with those obtained using Hill's estimation procedure used in previous works in the literature (see Sect. 3.2.2). The comparisons typically point out to similar conclusions for both estimation approaches.

sterling (GBP), and surprisingly, a decrease in the degree of heavy-tailedness of the Russian rouble (RUB).

These results have a number of implications. They underscore the need for robust econometric and statistical methods in the analysis of emerging country financial markets. Further, they highlight the aspects of macroeconomic management and policy in emerging countries. In a structural model that explains the determination of the tail indices of exchange rates, the tail indices of trading volumes as well as of sizes of market participants have a bearing (Gabaix et al. 2006). These reflect the extent of official intervention in the currency market (see the discussion in Sect. 3.2.3.) Further, estimates for emerging exchange rates may be used to forecast patterns in their future development and convergence to distributions with $\zeta \in (2, 4)$ as in the case of developed countries.

The section is organized as follows. Section 3.2.1 discusses the data on exchange rates of developed and emerging countries used in the analysis. Section 3.2.2 presents the empirical analysis of heavy-tailedness in the exchange rates considered. In Sect. 3.2.3, we discuss the implications of the empirical analysis and the results presented in Chap. 2 for several economic, financial, and econometric models, as well as for economic policy and forecasting, and make some suggestions for further research.

3.2.1   Data

We analyze daily exchange rates to US dollar (USD) for the currencies listed below, for the period from 1 January 1999 to 22 June 2012.[7] The developed country currencies we analyze in this section are: Australian dollar (AUD), CAD, CHF, Danish krone (DKK), Euro (EUR), GBP, JPY, Norwegian kroner (NOK), and Swedish Krona (SEK). The currencies of emerging countries in our analysis are: Chinese renminbi (CNY), Hong Kong dollar (HKD), Indian rupee (INR), KRW, MYR, RUB, Singapore dollar (SGD), Taiwan dollar (TWD), and THB. For an

---

[7]The exchange rate of Russian Ruble is available from the Central Bank of Russia, http://www.cbr. ru/eng/currency_base/default.aspx. The data source for all other exchange rates considered in this section is the Board of Governors of the Federal Reserve System, http://www.federalreserve.gov/ datadownload.

overview of exchange rate regimes in the emerging countries, see Patnaik et al. (2011) (Table 1 therein) and IMF AREAER, various issues.[8,9]

3.2.2 Estimation Results

Tables 3.1 and 3.2 present the estimation results for the tail indices in power law models (1.3) for all the exchange rates discussed in Sect. 3.2.1.

Due to sensitivity of the commonly used Hill's tail index inference approach to dependence and sample sizes used in estimation and the robustness of the log-log rank-size regression tail index estimation methods discussed in Sect. 3.1 (including favorable performance of the methods under deviations from power laws and heavy-tailed dependent GARCH processes that are often used for exchange rate modeling), we mainly focus on applications of the log-log rank-size regression approaches to inference on the foreign exchange rate tail indices. For illustration, we compare a number of the conclusions to those obtained using Hill's estimation approach (see Tables 3.1 and 3.2). The comparisons typically indicate similar conclusions for both the log-log rank-size regression and Hill's tail index estimation approaches.

Tables 3.1 and 3.2 report the tail index estimates $\hat{\zeta}_{RS}$ obtained from log-log rank-size regressions (3.5), with the optimal shift $\gamma = 1/2$ and the correct standard errors $\sqrt{\frac{2}{n}}\hat{\zeta}_{RS}$. The tables also provide the correct 95 % confidence intervals (3.6) for the true tail index values ζ. For comparison, the last three columns in Tables 3.1 and 3.2 provide Hill's approach estimates (3.2), their standard errors $\text{s.e.}_{Hill} = \frac{1}{\sqrt{n}}\hat{\zeta}_{Hill}$, and the corresponding 95 %-confidence intervals (3.3) for ζ. The estimates relate to the 5 and 10 % truncation levels for extreme observations as in (3.1): $n \approx 0.05N$ and $n \approx 0.1N$, where N is the total sample size for the time series.

Through the rest of this section, for brevity, we refer to the extreme observations used for estimation, defined with respect to the 5 and 10 % truncation levels in (3.1), by $AUD_{5\%}$, $AUD_{10\%}$, etc. Failure to reject the null hypothesis $H_a : \zeta = \zeta_0$ refers to the 5 % significance level and the two-sided alternative $H_a : \zeta \neq \zeta_0$. Rejection of H_0 refers to the 2.5 % significance level and the one-sided alternatives $H_a : \zeta < \zeta_0$

---

[8]The classification of the countries considered as emerging follows the *Economist*; this list includes Hong Kong, Singapore, and Saudi Arabia and the following economies in the Morgan Stanley Emerging Markets Index: Brazil, Chile, China (mainland), Colombia, Czech Republic, Egypt, Hungary, India, Indonesia, Iran, Israel, Jordan, Malaysia, Mexico, Morocco, Pakistan, Peru, Philippines, Poland, Russia, South Africa, South Korea, Taiwan, Thailand, Tunisia, Turkey, and Vietnam (the Morgan Stanley Capital International classifies the economies of Hong Kong and Singapore as developed countries).

[9]Two of these exchange rates, CNY and MYR, were pegged to the US dollar till 2005, and HKD regime is the related linked exchange (see Table 1 in Patnaik et al. 2011). Existence of peg periods for currencies, however, does not affect tail index estimates for their exchange rates since the estimates are based on largest absolute values of exchange rates (see Sect. 3.1).

Table 3.1 Tail index estimates for exchange rates in developed countries

Currency	Truncation (%)	$\hat{\zeta}_{RS}$	s.e.$_{RS} = \sqrt{\frac{2}{n}\hat{\zeta}_{RS}}$	95 % CI$_{RS}$, Eq. (3.6)	$\hat{\zeta}_{Hill}$	s.e.$_{Hill} = \sqrt{\frac{1}{n}\hat{\zeta}_{Hill}}$	95 % CI$_{Hill}$, Eq. (3.3)
AUD	10	2.80	0.22	(2.38, 3.23)	2.85	0.15	(2.54, 3.15)
	5	2.71	0.29	(2.14, 3.29)	2.88	0.22	(2.45, 3.32)
CAD	10	3.26	0.25	(2.77, 3.76)	2.99	0.16	(2.67, 3.30)
	5	3.46	0.37	(2.72, 4.19)	3.31	0.25	(2.81, 3.81)
CHF	10	3.92	0.30	(3.33, 4.51)	3.70	0.20	(3.31, 4.10)
	5	3.90	0.42	(3.07, 4.73)	4.14	0.32	(3.51, 4.76)
DKK	10	3.86	0.30	(3.28, 4.44)	3.64	0.20	(3.26, 4.03)
	5	3.98	0.43	(3.13, 4.83)	4.01	0.31	(3.40, 4.61)
EUR	10	4.23	0.33	(3.60, 4.87)	3.77	0.20	(3.37, 4.18)
	5	4.59	0.50	(3.62, 5.57)	4.28	0.33	(3.64, 4.92)
GBP	10	3.54	0.27	(3.01, 4.08)	3.65	0.20	(3.26, 4.04)
	5	3.34	0.36	(2.63, 4.05)	3.70	0.28	(3.14, 4.26)
JPY	10	3.39	0.26	(2.88, 3.90)	3.09	0.17	(2.76, 3.42)
	5	3.72	0.40	(2.93, 4.50)	3.54	0.27	(3.00, 4.07)
NOK	10	3.56	0.27	(3.03, 4.10)	3.24	0.18	(2.90, 3.59)
	5	3.90	0.42	(3.07, 4.73)	3.52	0.27	(2.99, 4.05)
SEK	10	3.64	0.28	(3.09, 4.19)	3.22	0.18	(2.88, 3.57)
	5	3.98	0.43	(3.13, 4.82)	3.67	0.28	(3.12, 4.22)

Note 1 January 1999 to 22 June 2012: $N = 3,390$, $10\%N = 339$, $5\%N = 170$

Table 3.2 Tail index estimates for exchange rates in emerging countries

Currency	Truncation (%)	$\hat{\zeta}_{RS}$	$s.e._{RS} = \sqrt{\frac{2}{n}}\hat{\zeta}_{RS}$	95 % CI$_{RS}$, Eq. (3.6)	$\hat{\zeta}_{Hill}$	$s.e._{Hill} = \sqrt{\frac{1}{n}}\hat{\zeta}_{Hill}$	95 % CI$_{Hill}$, Eq. (3.3)
CNY	10	2.18	0.17	(1.85, 2.51)	1.90	0.10	(1.70, 2.10)
	5	2.40	0.26	(1.89, 2.92)	2.36	0.18	(2.01, 2.72)
HKD	10	2.25	0.17	(1.91, 2.59)	1.91	0.10	(1.70, 2.11)
	5	2.57	0.28	(2.02, 3.11)	2.30	0.18	(1.95, 2.64)
INR	10	2.86	0.22	(2.43, 3.29)	2.52	0.14	(2.25, 2.78)
	5	3.16	0.34	(2.49, 3.83)	2.81	0.22	(2.39, 3.23)
KRW	10	2.32	0.18	(1.97, 2.66)	2.20	0.12	(1.97, 2.44)
	5	2.36	0.26	(1.86, 2.86)	2.23	0.17	(1.90, 2.57)
MYR	10	3.25	0.25	(2.76, 3.74)	2.56	0.14	(2.28, 2.83)
	5	4.08	0.44	(3.21, 4.94)	3.51	0.27	(2.98, 4.03)
RUB	10	2.41	0.18	(2.04, 2.77)	1.96	0.11	(1.75, 2.17)
	5	2.81	0.31	(2.22, 3.41)	2.57	0.20	(2.18, 2.95)
SGD	10	3.12	0.24	(2.65, 3.60)	2.84	0.15	(2.54, 3.14)
	5	3.36	0.36	(2.65, 4.07)	3.07	0.24	(2.61, 3.53)
THB	10	2.66	0.20	(2.26, 3.06)	2.37	0.13	(2.11, 2.62)
	5	2.85	0.31	(2.25, 3.46)	2.47	0.19	(2.10, 2.85)
TWD	10	2.53	0.19	(2.15, 2.91)	2.35	0.13	(2.10, 2.60)
	5	2.54	0.28	(2.00, 3.08)	2.77	0.21	(2.36, 3.19)

Note 1 January 1999 to 22 June 2012: $N = 3,390$, $10\%N = 339$, $5\%N = 170$

or $H_a : \zeta > \zeta_0$. Similar to the results in Tables 3.1–3.5, the confidence intervals discussed below are 95 % confidence intervals.

The results in Tables 3.1 and 3.2 point to remarkable differences in heavy-tailedness properties of the exchange rates of developed and emerging countries. The point estimates $\hat{\zeta}_{RS}$ for developed country exchange rates in Table 3.1 lie between 2.7 and 4.6. This range of values is preserved by estimates $\hat{\zeta}_{Hill}$, which lie between 2.8 and 4.3. These results are in line with the results for developed country financial markets, which report ζ estimates in the interval $(2, 5)$ for returns on stocks and stock indices.

The null hypothesis $\zeta = 2$ is rejected in favor of $\zeta > 2$ for all developed country exchange rates by both the log-log rank-size regression and Hill's estimation procedures (Table 3.1). The null hypothesis $\zeta = 3$ is not rejected for $AUD_{5\%}$, $AUD_{10\%}$, $CAD_{5\%}$, $CAD_{10\%}$, $GBP_{5\%}$, $JPY_{5\%}$, and $JPY_{10\%}$ using the log-log rank-size regression approach, and for $AUD_{5\%}$, $AUD_{10\%}$, $CAD_{5\%}$, $CAD_{10\%}$, $JPY_{10\%}$, $NOK_{5\%}$, $NOK_{10\%}$, and $SEK_{10\%}$ using Hill's estimation. Both approaches reject the null hypothesis $\zeta = 3$ in favor of $\zeta > 3$ for $CHF_{5\%}$, $CHF_{10\%}$, $DKK_{5\%}$, $DKK_{10\%}$, $EUR_{5\%}$, $EUR_{10\%}$, $GBP_{10\%}$ and $SEK_{5\%}$. The hypothesis is also rejected for $GBP_{5\%}$, and at the margin for $JPY_{5\%}$ using Hill's estimation, and for $NOK_{5\%}$, $NOK_{10\%}$ and $SEK_{10\%}$ using log-log rank-size regression. Both approaches do not reject the null hypothesis $\zeta = 4$ for $CHF_{5\%}$, $CHF_{10\%}$, $DKK_{5\%}$, $DKK_{10\%}$, $EUR_{5\%}$, $EUR_{10\%}$, $GBP_{5\%}$, $GBP_{10\%}$, $JPY_{5\%}$, $NOK_{5\%}$, and $SEK_{5\%}$; this hypothesis is also not rejected for $CAD_{5\%}$, $NOK_{10\%}$, and $SEK_{10\%}$ by the confidence intervals based on log-log rank-size regressions. Both methods reject the null hypothesis $\zeta = 4$ in favor of $\zeta < 4$ for $AUD_{5\%}$, $AUD_{10\%}$, $CAD_{10\%}$ and $JPY_{10\%}$. This hypothesis is also rejected in favor of $\zeta < 4$ for $CAD_{5\%}$, $NOK_{10\%}$, and $SEK_{10\%}$ using Hill's estimation.

The conclusions that can be drawn from Table 3.1 on the existence of the second, third, and fourth moments for the exchange rates of developed countries are similar for both estimation approaches. All developed country exchange rates have finite variances. In addition, CHF, DKK, EUR, and, apparently, GBP and SEK have finite third moments; however, the fourth moments of these exchange rates may be infinite. In contrast, according to both approaches, AUD and, apparently, CAD have infinite fourth moments, and both may have infinite third moments. The third and fourth moments may be infinite for JPY and NOK. All in all, the results in Table 3.1 show that CHF, DKK, and EUR exchange rates are less heavy-tailed than AUD, CAD, GBP, JPY, NOK and SEK exchange rates, with AUD and CAD being the most heavy-tailed. The latter set of exchange rates would appear to be subject to more extreme shocks.

The results in Table 3.2 for emerging countries exchange rates can be summarized as follows. The point estimates $\hat{\zeta}_{RS}$ lie between 2.1 and 4.1, and the point estimates $\hat{\zeta}_{Hill}$ lie between 1.9 and 3.5. In particular, for CNY, HKD, KRW, RUB, THB, and TWD, both the point estimates $\hat{\zeta}_{RS}$ and $\hat{\zeta}_{Hill}$ are less than 2.9. The null hypothesis $\zeta = 1$ is rejected in favor of $\zeta > 1$ for all emerging country exchange rates. The null hypothesis $H_0 : \zeta = 2$ is not rejected for $CNY_{5\%}$, $CNY_{10\%}$, $HKD_{10\%}$, $KRW_{5\%}$, and $KRW_{10\%}$ using the log-log rank-size regression approach, and for $CNY_{10\%}$, $HKD_{5\%}$, $HKD_{10\%}$, $KRW_{5\%}$, $KRW_{10\%}$, and $RUB_{10\%}$ using Hill's estimation. Both approaches reject the null hypothesis $H_0 : \zeta = 2$ in favor of $H_a : \zeta > 2$ for $INR_{5\%}$, $INR_{10\%}$, $MYR_{5\%}$, $MYR_{10\%}$, $RUB_{5\%}$, $SGD_{5\%}$, $SGD_{10\%}$, $THB_{5\%}$, $THB_{10\%}$, $TWD_{5\%}$, and $TWD_{10\%}$. The hypothesis is also rejected in favor of $H_a : \zeta > 2$ for $HKD_{5\%}$ and $RUB_{10\%}$ using the log-log rank-size regression approach and for $CNY_{5\%}$ using Hill's method. The log-log rank-size regression approach does not reject the hypothesis $H_0 : \zeta = 3$ for $HKD_{5\%}$, $INR_{5\%}$, $INR_{10\%}$, $MYR_{10\%}$, $RUB_{5\%}$, $SGD_{5\%}$, $SGD_{10\%}$, $THB_{5\%}$, $THB_{10\%}$, and $TWD_{5\%}$, while the hypothesis $H_0 : \zeta = 3$ is rejected in favor of $H_a : \zeta < 3$ for all other exchange rates in Table 3.2, except $MYR_{5\%}$, for which the hypothesis is rejected in favor of $H_a : \zeta > 3$. The hypothesis $H_0 : \zeta = 3$ is also not rejected for $INR_{5\%}$, $MYR_{5\%}$, $SGD_{5\%}$, $SGD_{10\%}$ and $TWD_{5\%}$ using Hill's estimation, which also leads to the rejection of the hypothesis $H_0 : \zeta = 3$ in favor of $H_a : \zeta < 3$ for all other exchange rates in Table 3.2. The null hypothesis $H_0 : \zeta = 4$ is not rejected only for $SGD_{5\%}$ using the log-log rank-size regression approach, and for $MYR_{5\%}$ using both the estimation methods; it is rejected in favor of $H_0 : \zeta < 4$ for all the other exchange rates.

The results in Table 3.2 thus imply that the first moments are finite for all emerging country exchange rates. The variance is finite for INR, MYR, SGD, THB, TWD and, apparently, for RUB. The variances may be infinite for CNY, HKD, and KRW. CNY and KRW have infinite third moments. The same conclusion holds, apparently, for HKD and RUB. The third moments may also be infinite for INR, MYR, SGD, THB, and TWD. The fourth moments may be infinite for MYR and are, apparently, infinite for SGD; they are infinite for all the remaining exchange rates in Table 3.2.

Summarizing the results in Tables 3.1 and 3.2, typically the tail indices of exchange rates in emerging countries are considerably smaller than those of developed economies. The heavy-tailedness properties of exchange rates of emerging countries are indeed more pronounced than those of their developed counterparts. A key difference is that the exchange rates of developed countries appear to have finite variances, in contrast to the exchange rates of several emerging countries. Similarly, the third moments are infinite, or may be infinite, for most of the emerging country exchange rates, while they are finite for most developed country exchange rates.

These differences can be described in terms of the maximal order p of their finite moments as follows (see Sect. 1.2): For $p = 2.7$, while the null hypothesis $\zeta = p$ is rejected in favor of $\zeta > p$ for all developed country exchange rates in Table 3.1 except AUD, it is not rejected or is rejected in favor of $\zeta < p$ for all the emerging country exchange rates in Table 3.2 except $MYR_{5\%}$. Tail index values $\zeta = p \in (2.6, 2.8)$ are in some sense at the dividing boundary between those characteristic of developed and emerging countries: while the moments of order $p \in (2.6, 2.8)$ are finite for most of the developed country exchange rates, they may be infinite, or are infinite, for most of the emerging country exchange rates.

Tables 3.3 and 3.4 present estimation results pertaining to the effects of the on-going economic and financial crisis on the heavy-tailedness of exchange rates. Developed country exchange rates appear to have become more pronouncedly heavy-tailed since the beginning of the crisis (Table 3.3). The crisis-period con-

Table 3.3 Tail index estimates for exchange rates in developed countries before and after the beginning of the on-going crisis in 2008

Currency	Truncation (%)	4 Jan. 1999 to 15 Sept. 2008			15 Sept. 2008 to 22 June 2012		
		$\hat{\zeta}_{RS}$	s.e.$_{RS} =$ $\sqrt{\frac{2}{n}}\hat{\zeta}_{RS}$	95 % CI$_{RS}$	$\hat{\zeta}_{RS}$	s.e.$_{RS} =$ $\sqrt{\frac{2}{n}}\hat{\zeta}_{RS}$	95 % CI$_{RS}$
AUD	10	3.91	0.35	(3.22, 4.61)	2.42	0.35	(1.73, 3.10)
	5	3.99	0.51	(2.99, 4.99)	2.54	0.52	(1.51, 3.57)
CAD	10	4.48	0.41	(3.69, 5.28)	3.19	0.46	(2.28, 4.10)
	5	5.02	0.64	(3.76, 6.28)	3.36	0.69	(2.00, 4.73)
CHF	10	4.89	0.44	(4.02, 5.75)	3.04	0.44	(2.18, 3.91)
	5	5.84	0.75	(4.37, 7.30)	2.78	0.57	(1.66, 3.90)
DKK	10	4.07	0.37	(3.34, 4.79)	3.84	0.56	(2.75, 4.93)
	5	3.84	0.56	(2.75, 4.93)	4.19	0.86	(2.49, 5.88)
EUR	10	4.97	0.45	(4.09, 5.85)	3.87	0.56	(2.77, 4.98)
	5	5.67	0.73	(4.25, 7.10)	4.24	0.87	(2.52, 5.95)
GBP	10	5.43	0.49	(4.47, 6.40)	2.86	0.42	(2.05, 3.67)
	5	6.28	0.80	(4.71, 7.86)	2.89	0.60	(1.72, 4.06)
JPY	10	3.80	0.34	(3.12, 4.47)	2.90	0.42	(2.07, 3.72)
	5	4.26	0.55	(3.19, 5.33)	3.37	0.70	(2.01, 4.73)
NOK	10	4.65	0.42	(3.83, 5.48)	3.54	0.51	(2.53, 4.54)
	5	5.49	0.70	(4.11, 6.86)	4.15	0.86	(2.47, 5.83)
SEK	10	4.76	0.43	(3.92, 5.61)	3.80	0.55	(2.72, 4.88)
	5	6.23	0.80	(4.67, 7.79)	3.75	0.77	(2.23, 5.26)

Note 1 Jan. 1999 to 15 Sept. 2008: $N = 2444, 10\%N = 244, 5\%N = 122$; 15 Sept. 2008 to 22 June 2012: $N = 947, 10\%N = 95, 5\%N = 47$

fidence intervals for the tail indices of $AUD_{10\%}$, $CHF_{5\%}$, $CHF_{10\%}$, $GBP_{5\%}$, and $GBP_{10\%}$ lie to the left of their pre-crisis confidence intervals. This points to structural breaks and statistically significant decreases in the tail indices of these exchange rates after the beginning of the crisis in 2008 that correspond to the increase in the degree of their heavy-tailedness and the likelihood of large fluctuations. The pre- and post-crisis confidence intervals in Table 3.3 for other currencies intersect implying that the tail indices of these currencies before and after the beginning of the crisis are statistically indistinguishable from each other. This is also true of emerging country currencies in Table 3.4 with two notable exceptions: for RUB, and for $MYR_{10\%}$, the post-crisis confidence intervals lie to right of the pre-crisis confidence intervals. Heavy-tailedness properties have become less pronounced for these currencies, suggesting a corresponding decrease in the likelihood of large fluctuations in their exchange rates.

In order to illustrate the appropriateness of the tail truncation levels (5 and 10 %) used in this section, we follow the analysis and suggestions in Embrechts et al. (1997) and Mikosch and Stărică (2000), and present the analogues of Hill's

Table 3.4 Tail index estimates for exchange rates in emerging countries before and after the beginning of the on-going crisis in 2008

Currency	Truncation (%)	4 Jan. 1999 to 15 Sept. 2008			15 Sept. 2008 to 22 June 2012		
		$\hat{\zeta}_{RS}$	s.e.$_{RS}$ = $\sqrt{\frac{2}{n}}\hat{\zeta}_{RS}$	95 % CI$_{RS}$	$\hat{\zeta}_{RS}$	s.e.$_{RS}$ = $\sqrt{\frac{2}{n}}\hat{\zeta}_{RS}$	95 % CI$_{RS}$
CNY	10	1.99	0.18	(1.64, 2.35)	2.46	0.36	(1.76, 3.15)
	5	2.20	0.28	(1.65, 2.76)	2.60	0.54	(1.55, 3.65)
HKD	10	2.05	0.19	(1.69, 2.42)	2.69	0.39	(1.93, 3.46)
	5	2.31	0.30	(1.73, 2.89)	2.89	0.60	(1.72, 4.06)
INR	10	2.45	0.22	(2.02, 2.89)	3.41	0.49	(2.44, 4.38)
	5	2.72	0.35	(2.04, 3.40)	3.83	0.79	(2.28, 5.37)
KRW	10	3.26	0.30	(2.68, 3.84)	2.26	0.33	(1.62, 2.90)
	5	3.93	0.50	(2.94, 4.92)	2.44	0.50	(1.45, 3.42)
MYR	10	2.28	0.21	(1.88, 2.69)	4.01	0.58	(2.87, 5.15)
	5	2.99	0.38	(2.24, 3.74)	4.82	0.99	(2.87, 6.77)
SGD	10	3.92	0.36	(3.23, 4.62)	3.14	0.46	(2.25, 4.03)
	5	4.10	0.52	(3.07, 5.12)	3.62	0.75	(2.15, 5.08)
THB	10	2.62	0.24	(2.15, 3.08)	4.07	0.59	(2.91, 5.22)
	5	2.95	0.38	(2.21, 3.69)	5.09	1.05	(3.03, 7.15)
RUB	10	1.87	0.17	(1.54, 2.20)	3.32	0.48	(2.38, 4.27)
	5	1.84	0.24	(1.38, 2.31)	4.27	0.88	(2.54, 6.00)
TWD	10	2.50	0.23	(2.06, 2.95)	2.46	0.36	(1.76, 3.16)
	5	2.61	0.33	(1.95, 3.26)	2.30	0.47	(1.37, 3.22)

Note 1 Jan. 1999 to 15 Sept. 2008: $N = 2444, 10\%N = 244, 5\%N = 122$; 15 Sept. 2008 to 22 June 2012: $N = 947, 10\%N = 95, 5\%N = 47$

plots for the log-log rank-size regression tail index estimates for EUR, GBP, and RUB (Figs. 3.1, 3.2 and 3.3). These are graphs of the log-log rank-size regression point estimates $\hat{\zeta}_{RS}$ of the tail indices for the currencies' exchange rates, for different values of the truncation levels n for extreme observations, together with the corresponding 95 %-confidence intervals 95 % CI$_{RS}$ in (3.6) for the true tail index values computed using log-log rank-size regressions. The figures highlight the relative stability of the point estimates across truncation levels. In particular, we note that the 95 % CI$_{RS}$ confidence intervals constructed for different tail truncation levels intersect. This shows that the tail indices in power law approximations of the tails of distributions of the exchange rates are statistically indistinguishable for different tail truncation levels. In addition, according to confidence intervals in Figs. 3.1, 3.2 and 3.3, the qualitative conclusions in this section on (in)finiteness of second, third, and fourth moments for the exchange rates remain unchanged regardless of the choice of truncation levels for extreme observations used in estimation.

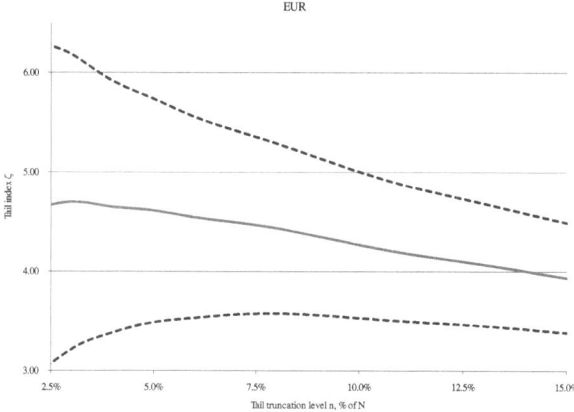

Fig. 3.1 Log-log rank-size estimates and 95 % confidence intervals for EUR tail index with different tail truncation levels

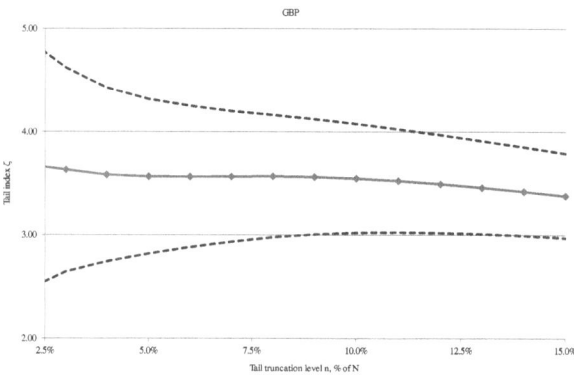

Fig. 3.2 Log-log rank-size estimates and 95 % confidence intervals for GBP tail index with different tail truncation levels

3.2.3   *Implications for Economic Models and Policy Decisions*

As discussed in Chap. 2, heavy-tailedness has crucial implications for the robustness of many economic and financial models, leading, in a number of settings, to reversals of conclusions drawn from them. The finding that the exchange rates of all countries considered in this section have tail indices greater than one is reassuring. The results in Sect. 2.1.3 show that the stylized facts on optimality of diversification are robust to heavy-tailedness of risks or returns in value at risk models as long as the distributions of these risks or returns are moderately heavy-tailed with tail indices $\zeta > 1$ and finite means. However, the stylized fact that portfolio diversification

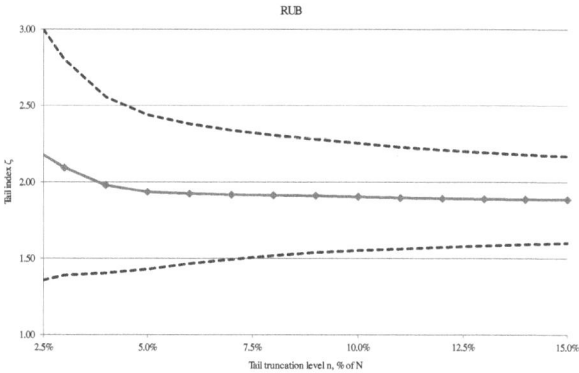

Fig. 3.3 Log-log rank-size estimates and 95 % confidence intervals for RUB tail index with different tail truncation levels

is preferable is reversed when tail indices are less than one and first moments are infinite. As discussed in Sect. 2.1.4, a similar conclusion holds for the efficiency property of the sample mean: the sample mean is the BLUE of the population mean in the sense of peakedness properties for moderately heavy-tailed populations with tail indices $\zeta > 1$. In addition, for such populations, the sample mean exhibits the property of monotone consistency, and thus, an increase in the sample size always improves its performance. However, the efficiency of the sample mean in the sense of its peakedness in estimating the population center decreases with sample size under extreme heavy-tailedness, when tail indices $\zeta < 1$.

Our estimates indicate that the tail indices may be less than two for exchange rates of several emerging countries. As discussed in Sect. 2.1.4, this presents a challenge to the applicability of standard statistical and econometric methods, including many classical approaches to inference based on variances and (auto)correlations, such as regression and spectral analysis, least squares methods, and autoregressive models. Our results imply that traditional econometric and statistical methods should be applied with care in studies of exchange rates for many emerging and developing countries. This is especially important in cases where tail indices are close to critical robustness boundary ($\zeta = 1$), or close to the threshold value for applicability of standard inference methods ($\zeta = 2$).

Our tail index estimates can further be used, together with estimates of the constant in log-log rank-size regressions and the implied estimates of the scaling constants in heavy-tailed models (1.3), for evaluation of commonly used risk measures such as the value at risk and expected shortfall relatively far in the tails of exchange rate distributions (see Sects. 1.2 and 2.1).

The results obtained are pertinent to economic policy making and macroeconomic forecasting. The finding that exchange rates of emerging and developing countries are typically more heavy-tailed than those of their developed counterparts, reflect their susceptibility to more frequent and extreme external and internal shocks. Secondly, estimated tail indices for the exchange rates of emerging and developing countries can be used to forecast the patterns in their future development, as they converge to distributions with tail indices $\zeta \in (2, 4)$ implied by empirical results and theoretical models in the literature for financial returns and exchange rates of developed country currencies (see Gabaix et al. 2006, Sect. 1.1 and the references therein). Third, our results suggest modification of structural models of heavy tailedness to suit emerging and developing countries exchange rates. For example, Gabaix et al. (2006) propose a model where the tail index value $\zeta \approx 3$ for financial returns in developed economies is generated by the trading actions of market participants who have a size distribution with tail indices $\zeta = 1$ (Zipf's law). Due to government intervention and regulation, it seems likely that the tail indices may be less than $\zeta = 1$ for largest participants in emerging and developing foreign exchange markets (implying deviations from the Zipf's law), with the states being some of the largest key traders. It is also likely that price impacts of trading volume differ between emerging and developed country currency markets. It will be interesting and important to provide estimation of these characteristics for different countries.

Developing structural models that explain the presence of heavy tails in emerging country exchange rates and the factors (including macroeconomic and institutional variables) that affect them would be an important direction for further research. This will require estimates of tail indices for size distributions of market participants, and of trading volume distributions. Cross-country analysis of relationships between tail index estimates and macro economic variables that are proximate to government intervention in foreign exchange markets should be interesting. Similar estimates for export and import volumes and their concentration across industries and trade partners are also important.

Application of other estimation and inference approaches, including the methods for adaptive selection of the tail sample fraction used in inference on heavy-tailedness (see Sect. 4.7 in Beirlant et al. 2004), small sample analogues of Hill's estimates developed in Huisman et al. (2001) and other robust econometric and statistical procedures under heavy tails such as the t-statistic based robust inference methods proposed in Ibragimov and Müller (2010) (see the next section) to this research issue would be of interest. Further analysis of structural breaks in the heavy-tailedness of exchange rates, using for example, tests developed in Quintos et al. (2001) will be useful. Going beyond the on-going economic and financial crisis, it would be of interest to study the impacts of changes in currency regimes and of other shocks (such as the 1998 Russian financial crisis).

3.3 Robust Inference Under Heterogeneity, Heavy Tails, and Dependence Using Conservativeness of Test Statistics

3.3.1 Description of the Robust Inference Approaches[10]

Focusing on the problem of estimation of the unknown mean of a stationary time series, Brillinger (1973) presents the important idea of dividing data into (say, q) groups, calculating the estimates of the mean (sample means) for each of the group and then basing the inference on the t-statistic of the resulting group estimates (sample means).[11] Brillinger (1973) further discusses (weak) dependence assumptions on the data generating and data sampling processes that guarantee that, asymptotically, the group sample means will be *independent normal* r.v.'s with *the same* variance (that is, *i.i.d.* and, thus, *homogeneous, normal* r.v.'s). Under such assumptions, the *asymptotic* distribution of the t-statistic of the group sample means used in inference is a *Student-t* distribution with $q-1$ degrees of freedom. Quantiles of the later distribution can thus be used in asymptotic hypothesis tests on the unknown mean and in constructing its confidence intervals.

Suppose now we want to do inference on some scalar parameter β of an economic or financial model in a large data set with n observations. For a wide range of models and estimators $\hat{\beta}$ of β, it is known that the distribution of $\hat{\beta}$ is approximately normal in large samples, that is, $\sqrt{n}(\hat{\beta} - \beta) \Rightarrow \mathcal{N}(0, \sigma^2)$ as $n \to \infty$, where "$\Rightarrow$" denotes convergence in distribution. Suppose further that the observations exhibit correlations of largely unknown form. If such correlations are pervasive and pronounced enough, then it will be very challenging to consistently estimate σ^2, and inference procedures for β that ignore the sampling variability of a candidate consistent estimator $\hat{\sigma}^2$ will have poor finite sample properties.

Ibragimov and Müller (2010) propose the following general approach to robust inference about an arbitrary parameter β under heterogeneity and correlation of a largely unknown form. The approach provides substantial extensions and refinements of Brillinger (1973) analysis of estimation the unknown mean of a stationary time series using the t-statistic in asymptotically *homogeneous* normal group sample means and its *limiting* Student-t distribution. These extensions cover, in particular, inference on an *arbitrary* unknown parameter of interest, possibly *nonstationary* models with *changing* or *stochastic* variances, and *heterogeneous* normal and *non-normal* limits for the parameter's group estimators, including many important settings with panel, clustered, spatially correlated and heavy-tailed data.

---

[10]This section has drawn upon material from Ibragimov and Müller (2010) "*t*-statistic based correlation and heterogeneity robust inference," *Journal of Business and Economic Statistics*, Vol. 28, No. 4, 453–468.

[11]We are very grateful to an anonymous reviewer of the book for the important reference to Brillinger (1973) that the authors of the manuscript and the paper this section is based on were unaware of.

The approach to robust inference under heterogeneity, correlation, and heavy-tailedness in Ibragimov and Müller (2010) is described as follows. Consider a partition of the original data set into $q \geq 2$ groups, with n_j observations in group j, and $\sum_{j=1}^{q} n_j = n$. Denote by $\hat{\beta}_j$ the estimator of β using observations in group j only. Suppose the groups are chosen such that $\sqrt{n}(\hat{\beta}_j - \beta) \Rightarrow \mathcal{N}(0, \sigma_j^2)$ for all j, and, also, such that $\sqrt{n}(\hat{\beta}_j - \beta)$ and $\sqrt{n}(\hat{\beta}_i - \beta)$ are asymptotically independent for $i \neq j$—this amounts to the convergence in distribution

$$\sqrt{n}(\hat{\beta}_1 - \beta, \ldots, \hat{\beta}_q - \beta)' \Rightarrow \mathcal{N}(0, \text{diag}(\sigma_1^2, \ldots, \sigma_q^2)), \quad \max_{1 \leq j \leq q} \sigma_j^2 > 0 \quad (3.7)$$

and $\{\sigma_j^2\}_{j=1}^{q}$ are, of course, unknown. The asymptotic Gaussianity of $\sqrt{n}(\hat{\beta}_j - \beta), j = 1, \ldots, q$, typically follows from the same reasoning as the asymptotic Gaussianity of the full sample estimator $\hat{\beta}$. The argument for an asymptotic independence of $\hat{\beta}_j$ and $\hat{\beta}_i$ for $i \neq j$, on the other hand, depends on the choice of groups and the details of the application (see Sect. 4 in Ibragimov and Müller 2010 for the discussion of such arguments for several common econometric models, including time series, panel, clustered and spatially correlated settings).

As discussed in Ibragimov and Müller (2010), one can perform an asymptotically valid test of level α, $\alpha \leq 0.05$ of $H_0 : \beta = \beta_0$ against $H_1 : \beta \neq \beta_0$ by rejecting H_0 when $|t_\beta|$ exceeds the $(1 - \alpha/2)$ percentile of the Student$-t$ distribution with $q - 1$ degrees of freedom, where t_β is the usual t-statistic

$$t_\beta = \sqrt{q} \frac{\overline{\hat{\beta}} - \beta_0}{s_{\hat{\beta}}} \quad (3.8)$$

with $\overline{\hat{\beta}} = q^{-1} \sum_{j=1}^{q} \hat{\beta}_j$, the sample mean of the group estimators $\hat{\beta}_j, j = 1, \ldots, q$, and $s_{\hat{\beta}}^2 = (q - 1)^{-1} \sum_{j=1}^{q} (\hat{\beta}_j - \overline{\hat{\beta}})^2$, the sample variance of $\hat{\beta}_j, j = 1, \ldots, q$.

In other words, the usual t-tests can be used in the presence of asymptotic heteroskedasticity in group estimators as long as the level of the tests is not greater than the typically used 5 % threshold.

As discussed in Ibragimov and Müller (2010), the t-statistic approach provides a number of important advantages over the existing methods. In particular, it can be employed when data are potentially heterogeneous and correlated in a largely unknown way. In addition, the approach is simple to implement and does not need new tables of critical values. The assumptions of asymptotic normality for group estimators in the approach are explicit and easy to interpret, in contrast to conditions that imply validity of alternative procedures. Furthermore, as shown in Ibragimov and Müller (2010), the t-statistic based approach to robust inference efficiently exploits the information contained in these regularity assumptions, both in the small sample settings (uniformly most powerful scale invariant test against a benchmark alternative with equal variances) and also in the asymptotic frameworks.

It is important to emphasize that the asymptotic efficiency results for t-statistic based robust inference further imply that it is not possible to use data dependent methods to determine the optimal number of groups q to be used in the approach when the only assumption imposed on the data generating process is that of asymptotic normality for the group estimators $\hat{\beta}_j$. The numerical results presented in Ibragimov and Müller (2010) demonstrate that, for many dependence and heterogeneity settings considered in the literature and typically encountered in practice for time series, panel, clustered and spatially correlated data, the choice $q = 8$ or $q = 16$ leads to robust tests with attractive finite sample performance.

It is further important to note that the t-statistic approach described provides a formal justification for the widespread Fama–MacBeth method for inference in panel regressions with heteroskedasticity (see Fama and MacBeth 1973). In the approach, one estimates the regression separately for each year, and then tests hypotheses about the coefficient of interest using the t-statistic of the resulting yearly coefficient estimates. The Fama–MacBeth approach is a special case of the t-statistic based approach to inference, with observations of the same year collected in a group.

In addition, the same approach remains valid under deviations from normality as in the case of heavy-tailed models, as long as the estimators $\hat{\beta}_j, j = 1, \ldots, q$, are asymptotically independent and converge (at an arbitrary rate) to heavy-tailed scale mixtures of normals. Namely, the approach is asymptotically valid if

$$\{m_n(\hat{\beta}_j - \beta)\}_{j=1}^q \Rightarrow \{Z_j V_j\}_{j=1}^q \tag{3.9}$$

for some real sequence m_n, where $Z_j \sim i.i.d.\ \mathcal{N}(0, 1)$, the random vector $\{V_j\}_{j=1}^q$ is independent of the vector $\{Z_j\}_{j=1}^q$ and $\max_j |V_j| > 0$ almost surely (see the discussion and the numerical results on performance of the approach in heavy-tailed models in Sect. 3.3.3.). The class of limiting scale mixtures of normals in (3.9) is a rather large class of distributions: it includes, for instance, Student$-t$ distributions with arbitrary numbers of freedom d that follow power laws (1.1)–(1.3) with the tail indices $\zeta_1 = \zeta_2 = \zeta = d$ (including the Cauchy distribution with $\zeta = d = 1$), double exponential distributions, logistic distributions and all symmetric stable distributions (see Sect. 2.1.2) that typically arise as distributional limits of estimators in econometric models under heavy-tailedness with infinite variances.

3.3.2 Small Sample Properties of t-Statistics and Inequality Indices Based on Self-Normalized Sums

The robust approach to asymptotic inference in Ibragimov and Müller (2010) reviewed in the previous section relies on the following powerful result on small sample properties of the t-statistic in heterogeneous normal observations due to Bakirov and Székely (2005).

Let $X_j, j = 1, \ldots, q$, with $q \geq 2$, be independent Gaussian random variables with common mean $E[X_j] = \mu$ and variances $V[X_j] = \sigma_j^2$. Consider the usual t-statistic for the hypothesis test $H_0 : \mu = 0$ against the alternative $H_a : \mu \neq 0$:

$$t = \sqrt{q}\frac{\bar{X}}{s_X}. \tag{3.10}$$

If the variances σ_j^2 are the same: $\sigma_j^2 = \sigma^2$ for all j, by definition, the critical value cv of $|t|$ is given by the appropriate percentile of the distribution of a Student–t distributed random variable T_{q-1} with $q - 1$ degrees of freedom.

The case of equal variances is extremal for the t-statistic in (1) in the following sense (see the results in Bakirov and Székely 2005). Let $cv_q(\alpha)$ be the critical value of the usual two-sided t-test of H_0 against H_a of level $\alpha \leq 0.05 : P(|T_{q-1}| > cv_q(\alpha)) = \alpha$. Then then for all $q \geq 2$,

$$\sup_{\{\sigma_1^2,\ldots,\sigma_q^2\}} P(|t| > cv_q(\alpha)|H_0) = P(|T_{q-1}| > cv_q(\alpha)) = \alpha. \tag{3.11}$$

The conservativeness result in (3.11) does not hold for 10 % level with $\alpha = 0.1$.

The conservativeness properties of t-statistic given by (3.11) imply analogous results for the tail probabilities of self-normalized sums

$$S_q = \sum_{j=1}^{q} X_j / \left(\sum_{j=1}^{q} X_j^2 \right)^{1/2} \tag{3.12}$$

and their squares using the equality

$$P(|t| > y) = P\left(S_q^2 > \frac{qy^2}{y^2 + q - 1} \right) \tag{3.13}$$

for all $y > 0$.

As follows from Bakirov (1998) and the recent paper by Ibragimov and Müller (2013), conservativeness properties similar to those for one-sample t-statistics also hold for two-sample t-statistics (Behrens–Fisher statistics) for testing equality of means: that is, for commonly used significance levels, the two-sample t-tests remain conservative for underlying observations that are independent and Gaussian with heterogeneous variances. These small sample conservativeness results provide the basis for the development of new approaches to asymptotic robust inference under heterogeneity, dependence, and heavy-tailedness using two-sample t-statistics in Ibragimov and Müller (2013) and their applications in several important problems including robust tests for structural breaks and the analysis of treatment effects. Similar to the t-statistic based approach discussed in the previous, the large sample inference using two-sample t-statistics can be conducted as follows: partition the data into some number of groups, estimate the model for each group, and conduct the standard two-sample t-test on equality of parameters (e.g., the test of the no-break hypothesis of equality of pre- and post-break parameters or the test on equality of parameters in the treatment and no-treatment groups) with the resulting parameter estimators of interest.

The above small sample conservativeness properties of t-statistics are closely related to the properties of several income inequality measures and other economic indicators such as the coefficient of variation.

Consider a sample of observations $X_1, \ldots, X_q$, $q \geq 2$ (e.g., income or wealth levels of q individuals). Let $cv_q(\alpha)$ denote the $(1 - \alpha/2)$–quantile of Student-t distribution with $(q - 1)$ degrees of freedom: $P(|T_{q-1}| > cv_q(\alpha)) = \alpha$.

Representations similar to t-statistic $t = \sqrt{q}\overline{X}/s_X$, and self-normalized sums $S_q = \frac{\sum_{j=1}^{q} X_j}{\sqrt{\sum_{j=1}^{q} X_j^2}}$ in (3.10) and (3.12) hold for a number of variables of interest in economics and finance, including, for instance, one of the widely used inequality measures, the empirical coefficient of variation $\widehat{CV}_X = s_X/\overline{X} = \sqrt{q}/t$.[12] These representations together with the conservativeness results for t-statistics and self-normalized sums given by (3.13) and (3.11) imply similar results for the tail probabilities of the empirical coefficient of variation $\widehat{CV}_X$, and a number of other variables in economics and finance. These comparisons for the empirical inequality measures such as $\widehat{CV}_X$ and their analogues for transformations (such as logarithms) of the observations X_j provide comparisons between the tail probabilities and the cdf's of these measures under heterogeneity and heavy-tailedness and those in the standard homogeneous Gaussian case.

Below, $Y_j = \log X_j$ denote the logarithms of the observations, provided that $X_j > 0$. In addition, let $Z_1, \ldots, Z_q$ denote the i.i.d. standard normal r.v.'s: $Z_j \sim \mathcal{N}(0, 1)$ and let $V_j = e^{Z_j}$, $j = 1, .., q$, be the corresponding homogeneous log-normal r.v.'s. Following the above notation, in Proposition 3.3.1, $\widehat{CV}_Y = s_Y/\overline{Y}$ and $\widehat{CV}_Z = s_Z/\overline{Z}$.

Proposition 3.3.1 *If $Y_1 = \log X_1, \ldots, Y_q = \log X_q$ are independent heterogeneous normal r.v.'s $Y_j \sim \mathcal{N}(0, \sigma_j^2)$ or are scale mixtures of normals (for instance, independent not necessarily identically distributed stable r.v.'s), then*

$$P(0 < \widehat{CV}_Y < y) \leq P(0 < \widehat{CV}_Z < y) \tag{3.14}$$

$$P(|\widehat{CV}_Y| < y) \leq P(|\widehat{CV}_Z| < y) \tag{3.15}$$

---

[12]Similar representations also hold for the estimators of Sharpe ratio $\widehat{SR}$ for excess returns X_j, $j = 1, \ldots, q$. In addition, this is the case for the Herfindahl–Hirschman Index of market concentration that has the form $HHI = \sum_{j=1}^{q} X_j^2 / \left(\sum_{j=1}^{q} X_j \right)^2$ and is, thus, the inverse of the square of the self-normalized ratio in (3.12) for firm sizes X_j, $j = 1, \ldots, q$. The representations also hold, for instance, for commonly used sample split prediction test statistics employed in testing for time series stationarity (see Loretan and Phillips 1994a,b, and references therein).

for all $y < 1/(cv_{q-1}(0.05)\sqrt{q})$. In general, inequalities (3.14), (3.15) do not hold for $y < 1/(cv_{q-1}(0.1)\sqrt{q})$.

Inequalities (3.14) and (3.15) in Proposition 3.3.1 provide first order stochastic dominance comparisons of inequality measured by the coefficient of variation for log-incomes. The comparisons are between the income distribution $(X_1, \ldots, X_q)$ generated by heterogeneous log-normal X_j and the income distribution $(V_1, \ldots, V_q)$ generated by the homogeneous log-normal V_j. Naturally, inequality is expected to be less in the latter case of homogeneity. Inequalities (3.14) and (3.15) imply that homogeneity is indeed likely to reduce inequality, as measured by the coefficient of variation for log-income, in the region of small values of this inequality measure. However, according to Proposition 3.3.1, this does not hold in general. This conclusion may be viewed as an indicator that the coefficient of variation is a poor measure of inequality for some parts of the income or wealth distribution, including the middle and high income and wealth ranges.

3.3.3 Robust Inference in Heavy-Tailed Models[13]

Many works in econometrics and statistics have focused on the analysis of inference for the mean EX_t or, more generally, the location parameter of stationary heavy-tailed time series $\{X_t\}_{t=1}^{\infty}$ with infinite variance. As is illustrated by the results in Sect. 2.6, robust inference in such problem is crucial in a number of settings in economics and finance, including the analysis of properties of important models of firm growth.

Let $\{\epsilon_t\}_{t=-\infty}^{\infty}$ be a sequence of i.i.d. random variables that have a distribution with regularly varying tails in the form of power laws (1.1)–(1.3)

$$P(\epsilon_t < -x) \sim x^{-\zeta}L(x), \quad P(\epsilon_t > x) \sim x^{-\zeta}L(x), \quad x \to \infty, \tag{3.16}$$

where $\zeta \in (1, 2)$ and $L(x)$ is a slowly varying (at ∞) function, that is, $\lim_{x\to\infty} L(tx)/L(x) = 1$ for all $t > 0$ (note that this assumption concerns only the tails of the distribution of $\epsilon_t's$; it is not assumed that the random variables have a symmetric distribution).

According to the results in Logan et al. (1973) (see also Loretan and Phillips 1994a; Phillips 1990; Phillips and Hajivassiliou 1987), the t-statistics of i.i.d. random variables $\{\epsilon_t\}$ have a well-defined asymptotic distribution. This asymptotic distribution, however, depends in a complicated way on the tail index ζ of ϵ_t's that makes problematic the use of the latter weak convergence results, with further substantial complications under dependence.

---

[13]This section has drawn upon the extended version of supplementary material to Ibragimov and Müller (2010).

Motivated by the above problems and also by the failure of the bootstrap for the sample mean of heavy-tailed observations, several papers in statistics and econometrics have focused on developing subsampling inference for the mean of heavy-tailed and possibly dependent time series (see Politis et al. 1999; Romano and Wolf 1999 for subsampling methods for i.i.d. thick-tailed observations, McElroy and Politis 2002 for the case of linear processes and Kokoszka and Wolf 2004 for extensions to GARCH-like time series models). In particular, McElroy and Politis (2002) show that the t-statistic of stationary linear processes

$$X_t = \sum_{j=-\infty}^{\infty} c_j \epsilon_{t-j}, \quad \sum_{j=-\infty}^{\infty} |c_j| < \infty, \quad t \geq 1, \tag{3.17}$$

driven by heavy-tailed i.i.d. innovations ϵ_t satisfying (3.16) have a well-defined limit distribution that depends on the filter coefficients $\{c_j\}_{j=-\infty}^{\infty}$ as well as on the tail index ζ of ϵ_t's. McElroy and Politis (2002) further use this result to show that subsampling leads to asymptotically valid t-statistic based inference on the mean EX_t under the additional assumption that the linear time series $\{X_t\}$ is strong mixing and derive asymptotically correct subsampling confidence intervals for the mean of such processes without knowledge or explicit estimation of their tail index or dependence parameters.[14]

As discussed in McElroy and Politis (2002) (see also Politis et al. 1999, Chap. 11, and Romano and Wolf 1999), subsampling is the only feasible method available in the literature for solving the problem that the limiting distributions of sums and t-statistics of i.i.d. and autocorrelated time series are not pivotal. This is due to the complicated form of the asymptotic distributions and the difficulties in inference on the tail index ζ and the rate of convergence of its estimators needed in the alternative methods.

As discussed in the previous section, the fact that scale mixtures of normals include, as particular cases, a number of heavy-tailed distributions such as symmetric stable and Student-t, makes the t-statistic based approach to robust inference discussed in this section applicable in the statistical analysis under thick-tailedness and dependence.[15] In particular, the t-statistic based inference considered in Sect. 3.3.1 provides an alternative to subsampling methods for the mean of heavy-tailed autocorrelated time series. Similar to subsampling methods, the t-statistic based robust inference approach does not require knowledge or explicit estimation of

---

[14]The strong mixing assumption is satisfied, in particular, for MA processes of finite order and stationary AR(1) models driven by absolutely continuous i.i.d. innovations.

[15]In addition to the statistical analysis of location parameter of heavy-tailed time series considered in this section, further applications of the t-statistic based robust inference approach that are currently under way by the authors of the book and their co-authors include inference on income and wealth inequality measures and inference on autocorrelation functions of heavy-tailed GARCH and related processes where application of standard asymptotic results becomes problematic (see the discussion in Sects. 1.2, 2.1.4 and 2.7, and the references therein).

the tail index or dependence characteristics of thick-tailed processes. Furthermore, in contrast to subsampling, the validity of the t-statistic based inference does not require that the time series in consideration satisfy strong mixing conditions that are usually difficult to check (see Doukhan 1994 and the discussion in Nze and Doukhan 2004). The applications of the proposed method hold in all settings satisfying condition (3.9) that also includes models beyond heavy-tailed ones.

Let $\{X_t\}_{t=1}^{\infty}$ be a linear process in (3.17) driven by i.i.d. innovations $\{\epsilon_t\}_{t=-\infty}^{\infty}$ satisfying (3.16), and let $\mu = EX_t$. As in Brillinger (1973) and Ibragimov and Müller (2010) discussion of time series applications of the t-statistic based approach to robust inference (see Sect. 3.1 in the latter paper), consider dividing the sample $\{X_t\}_{t=1}^{T}$ into q (approximately) equal sized groups of consecutive observations: the observation indexed by t, $t = 1, \ldots, T$, is element of group j if $t \in \mathcal{I}_j = \{s : (j-1)T/q < s < jT/q\}$ for $j = 1, \ldots, q$.

For some positive sequence $d_T \to \infty$ (e.g., $d_T = T^{1/\zeta}$ if $L(x) = 1$ in (3.16)), the finite-dimensional distributions of the partial sum process $d_T^{-1} \sum_{t=1}^{[Tr]} (X_t - \mu)$ converge weakly to those of a symmetric Lévy ζ−stable process $S_\zeta(r)$ (see Astrauskas 1983; Davis and Resnick 1985, Lemma 1 in Avram and Taqqu 1992 and Remarks 3.20 in Phillips and Solo 1992).[16] Therefore, as in (3.9), the following joint convergence holds for the sample means $\overline{X}^{(j)}$ of observations in groups $j = 1, \ldots, q$:

$$Td_T^{-1}\left(\overline{X}^{(j)} - \mu\right) \to S_\zeta^j \tag{3.18}$$

as $T \to \infty$, where S_ζ^j are i.i.d. symmetric ζ−stable random variables (the increments of the stochastic process $S_\zeta(r)$ multiplied by $q^{1/\zeta-1}$). From the discussion in Sect. 3.3.1 we thus conclude that one can perform an asymptotically valid test of level α, $\alpha \leq 0.05$ of $H_0 : \mu = \mu_0$ against $H_1 : \mu \neq \mu_0$ by rejecting H_0 when $|t_\mu|$ exceeds the $(1 - \alpha/2)$ percentile of the student−t distribution with $q - 1$ degrees of freedom, where t_μ is the t-statistic calculated using the sample means $\overline{X}^{(j)}$:

$$t_\mu = \sqrt{q}(\tfrac{1}{q}\textstyle\sum_{j=1}^{q} \overline{X}^{(j)} - \mu_0)/s_q, \; s_q^2 = (q-1)^{-1}\sum_{j=1}^{q}\left(\overline{X}^{(j)} - \tfrac{1}{q}\sum_{j=1}^{q}\overline{X}^{(j)}\right)^2.^{17}$$

Table 3.1 presents the results of the Monte Carlo comparisons of small sample properties of subsampling and the t-statistic based inference for the mean of heavy-tailed dependent time series. As in McElroy and Politis (2002), we provide the results for the AR(1) model with the autoregressive parameter 0.5 and an MA(11) process for the sample sizes $T = 100, 1000$ and the tail index ζ equal to 1.2 and 1.5.

---

[16]It is important to note that, despite this convergence of finite-dimensional distributions, the process $d_T^{-1} \sum_{t=1}^{[Tr]} (X_t - \mu)$ does not converge weakly to $S_\zeta(r)$ in $\mathbb{D}[0, 1]$ endowed with the usual Skorohod topology (see Avram and Taqqu 1992 and Remarks 3.20 in Phillips and Solo 1992).

[17]The fact that, in this approach, the inference is based on the t-statistic for block *sample means* is in contrast to subsampling that uses the empirical distribution function of t-*statistics* calculated over the block of size b as an approximation to the limit distribution of the full-sample t-statistic $t_T = \sqrt{T}(\overline{X}_T - \mu)/s_T$.

The AR(1) model is $X_t = 0.5X_{t-1} + \epsilon_t$, $t = 1, \ldots, T$, with $\rho = 0.5$ and $\rho = 0.9$, $X_0 = \epsilon_0$ and symmetric ζ—stable i.i.d. random variables ϵ_t, $t \geq 0$. The MA(11) time series is $X_t = \sum_{j=1}^{10} \psi_j \epsilon_{t-j}$, $t = 1, \ldots, T$, with the coefficients $\{\psi_j\}$ equal to 0.03, 0.05, 0.07, 0.10, 0.15, 0.20, 0.15, 0.10, 0.07, 0.05, 0.03. The choice of the (fixed) block sizes b for subsampling also follows that in McElroy and Politis (2002).[18]

The results in Table 3.5 show that the t-statistic based approach to inference on the mean of the considered weakly dependent time series with heavy tails (3.16) substantially outperforms subsampling, both in terms of controlling size and with respect to the small sample power properties[19] Unreported results indicate that, overwhelmingly, this is also the case for other values of the tail index, the autoregressive parameters in the AR(1) specifications and other choices of the subsampling block size b. The dominance of the t-statistic based inference for the mean over subsampling is especially pronounced for the AR(1) heavy-tailed processes with the tail index ζ close to 1, with the empirical rejection probabilities being, e.g., for $\zeta = 1.2$ and $T = 1,000$, 63.1 % for the former approach with $q=16$ (the corresponding empirical size is equal to 3 %) vs. 5.2 % for the latter method with $b = 2$ (the corresponding empirical size is 2.9 %). The advantages of the t-statistic based inference on the mean over subsampling are also very significant in the case of the MA(11) model, both in the case of small (e.g., $\zeta = 1.2$) and large (e.g., $\zeta = 1.8$) values of the tail index ζ; the advantages further continue to exist in the case of AR(1) processes with $\zeta's$ close to two, including the case $\zeta = 1.8$ reported in Table 3.1.

In addition, as is seen from the numerical results presented below, the dependence of finite-sample performance of the t-statistic based inference on the parameter q (the number of blocks) is much weaker than the dependence of subsampling performance on the block size b. As discussed in McElroy and Politis (2002), while some insights on optimally picking b are available in the case of finite second moments (see Politis et al. 1999), little is known on the problem in the presence of dependence and heavy-tailedness. For example, even the computationally expensive approaches to the optimal choice of the block size b for i.i.d. heavy-tailed observations considered in Romano and Wolf (1999) and Politis et al. (1999), Chaps. 9 and

---

[18]We could not replicate several numerical results for subsampling reported in McElroy and Politis (2002), in part because the formulas that enter the description of subsampling inference therein break down for the block size $b = 1$ present in the tables in this section. We have noticed that Tables 1 and 2 in McElroy and Politis (2002) for AR(1) processes with autoregressive parameters 0.5 and 0.9 are identical up to Monte Carlo. Also, it is our sense that the values of the block sizes b in the first column of tables in McElroy and Politis (2002) do not always correspond to the actual values used in the computations.

[19]One should note that subsampling is generally designed to work under broad assumptions and hence sometimes it does not perform as well as more highly tuned methods. In the context of this section, subsampling also works in the case of unbalanced tails of the innovations ϵ_t with $\lim_{x\to\infty} P(\epsilon_t > x)/P(\epsilon_t < -x) = c \neq 1$ (see McElroy and Politis 2002). Balancedness of the tails of ϵ_t (the property that the above limit $c = 1$ for ϵ_t in (3.16)) that implies that the limits in (3.18) are *symmetric* stable r.v.'s (and, thus, scale mixtures of normals as in (3.9)) is important for applications of the t-statistic based inference in the analysis dealt with.

Table 3.5 Small sample results in a time series location model with symmetric ζ-stable disturbances

				t-statistic (q)				Subsampled t-statistic (b)			
				2	4	8	16	2	4	8	16
DGP	ζ	T	μ	Size							
MA	1.2	100	0	3.8	3.4	4.7	9.8	0.0	0.5	3.4	18.7
MA	1.2	1,000	0	3.6	3.1	2.9	3.1	2.0	3.3	7.1	16.5
MA	1.8	100	0	4.9	5.0	6.7	12.1	0.0	0.1	1.6	9.4
MA	1.8	1,000	0	4.9	4.7	5.0	5.0	0.1	0.1	0.2	0.9
AR	1.2	100	0	3.9	3.2	3.7	5.4	0.2	7.0	18.6	32.6
AR	1.2	1,000	0	4.1	3.0	2.5	3.0	2.9	10.5	20.5	26.3
AR	1.8	100	0	4.7	4.4	5.3	6.9	0.0	1.1	6.5	14.9
AR	1.8	1,000	0	4.7	4.6	4.8	4.8	0.1	0.3	1.9	4.8
DGP	ζ	T	μ	Non-size adjusted power							
MA	1.2	100	1.0	10.4	30.1	45.3	56.4	0.6	10.5	28.6	49.4
MA	1.2	1,000	1.0	14.2	45.7	58.9	63.8	2.5	19.0	44.1	59.1
MA	1.8	100	0.4	14.2	38.4	58.9	73.5	0.0	2.9	20.6	49.2
MA	1.8	1,000	0.2	19.0	56.8	76.0	82.0	0.1	0.1	15.8	51.1
AR	1.2	100	2.0	10.8	29.2	42.5	50.5	3.5	36.2	51.3	60.2
AR	1.2	1,000	2.0	14.9	45.0	58.3	63.1	5.2	52.3	65.7	69.5
AR	1.8	100	0.8	14.2	37.7	55.4	65.6	0.1	24.1	51.2	66.4
AR	1.8	1,000	0.4	18.7	57.2	75.6	81.3	0.1	34.6	70.0	78.6

Notes: Rejection probabilities of nominal 5 % level two-sided tests about μ in the model $y_t = \mu + u_t$, $t = 1, \ldots, T$, where $u_t = 0.5u_{t-1} + S_t$ and $u_0 = S_0$ (AR) and $u_t = \sum_{j=0}^{10} \psi_j S_{t-j}$ with $\{\psi_j\}_{j=0}^{10} = \{0.03, 0.05, 0.07, 0.1, 0.15, 0.2, 0.15, 0.1, 0.07, 0.05, 0.03\}$ (MA), and S_t are i.i.d. ζ-symmetric stable distributed. The subsampled t-statistic rejects if the full sample OLS t-statistic falls outside the 2.5 and 97.5 % quantiles of the empirical distribution function of OLS t-statistics computed on all $T - b + 1$ consecutive subsamples of length b, as described in detail in McElroy and Politis (2002). Based on 10,000 replications

11, are heuristic in nature and are not known to satisfy any asymptotic optimality properties. In addition, if tail index ζ is close to 1, the methods for the block size selection available in the literature result in the empirical coverage probabilities of subsampling confidence intervals for the mean that are quite far apart from the nominal coverage levels both in the i.i.d. (Politis et al. 1999, Chaps. 9) and the dependent case (Kokoszka and Wolf 2004).

3.4   Conclusion

As discussed in this book, the presence of heavy-tailed risk distributions in economics, finance, and insurance has important implications: it decreases the potential for diversification, which in turn affects the welfare of risk averse market participants, may magnify agency problems, and under some circumstances even lead to the break down of markets for risk. These implications are well understood and, as noted in the introduction, the study if heavy-tailedness is in some ways mature, dating back 50 years. In other ways, however, it is still in its infancy. We mention two areas where we believe future research may deepen our understanding significantly.

First, previous research—including ours—has usually focused on the effects of heavy-tailed shocks on economic, financial, or insurance markets, by modeling these shocks as exogenous. There is substantial evidence, however, that many of these shocks seem to arise endogenously, "within the system." This certainly applies to the recent financial crisis, which is commonly referred to as "systemic," the term having exactly this interpretation. It can also be argued that it is also the case for many other events, e.g., the stock market crash of October 19, 1987, which was not triggered by any public news events. Similarly, shocks to individual firms, e.g., operational risk events, may be triggered by the behavior of agents within the system, e.g., when the interaction between a firm and its customers leads to major law-suits. An exciting area of future research, with potentially significant policy implications, would focus on furthering our understanding of how the joint strategic actions of agents in an economic system endogenously may generate heavy-tailed risks.

Second, the results on (non-)robustness of economic and financial models presented in the book motivate the development and applications of robust inference approaches under heavy tails, heterogeneity, and dependence. The inference methods discussed in the book include the log-log rank-size regression approaches to tail index estimation with correct standard errors and optimal shifts in ranks and t-statistic-based approaches to robust inference under heterogeneity, dependence, and heavy-tailedness of largely unknown type. The inference methods discussed are illustrated by several empirical applications, including the analysis of heavy-tailedness properties of foreign exchange markets in emerging and developed countries. The approaches dealt with in this chapter can also be used in the analysis of heavy-tailedness properties of many other economic and financial variables and the effects of crises on them (see Ibragimov and Ibragimov 2014a for the empirical analysis of heavy-tailedness of income and wealth distributions and upper tail income and wealth inequality in Russia and the world and Ibragimov and Ibragimov 2014b for the study of the effects of the 2008 crisis on the dynamics of unemployment, economic growth and other key variables and indicators in CIS economies and world markets). Further applications of the presented inference methods and related robust inference approaches in the analysis of a number of important problems in economics and finance complicated by heavy-tailedness, heterogeneity, and dependence are currently under way by the authors and their co-authors.

Bibliography

Abramowitz, M., & Stegun, I. A. (1970). *Handbook of Mathematical Functions*, New York: Dover.

Acemoglu, D., Carvalho, V. M., Ozdaglar, A., & Tahbaz-Saleh, A. (2012) . The network origins of aggregate fluctuations. *Econometrica, 80*, 1977–2016.

Acharya, V. (2009). A theory of systemic risk and design of prudent bank regulation. *Journal of Financial Stability, 5*, 224–255.

Adams, W. J., & Yellen, J. L. (1976). Commodity bundling and the burden of monopoly. *Quarterly Journal of Economics, 90*, 475–498.

Adrian, T., & Brunnermeier, M. (2011). CoVar. Federal Reserve Bank of New York Staff Report no. 348. Available at http://www.newyorkfed.org/research/staff_reports/sr348.pdf.

Agarwal, R., & Gort, M. (1996). The evolution of markets and entry, exit and survival of firms. *Review of Economics and Statistics, 78*, 489–498.

Akgiray, V., Booth, G. G., & Seifert, B. (1988). Distribution properties of Latin American black market exchange rates. *Journal of International Money and Finance, 7*, 37–48.

Allen, F., & Gale, D. (2000). Financial contagion. *Journal of Political Economy, 108*, 1–33.

An, M. Y. (1998). Logconcavity versus logconvexity: A complete characterization. *Journal of Economic Theory 80*, 350–369.

Andrews, D. W. K. (1993). Tests for parameter instability and structural change with unknown change point. *Econometrica, 61*(4), 821–856. http://dx.doi.org/10.2307/2951764.

Artzner, P., Delbaen, F., Eber, J.-M., & Heath, D. (1999). Coherent measures of risk. *Mathematical Finance, 9*, 203–228.

Astrauskas, A. (1983). Limit theorems for sums of linearly generated random variables. *Lithuanian Mathematical Journal, 23*, 127–134.

Audretsch, D. B. (1991). New firm survival and the technological regime. *Review of Economics and Statistics, 73*, 441–450.

Avram, F., & Taqqu, M. S. (1992). Weak convergence of sums of moving averages in the α-stable domain of attraction. *Annals of Probability, 20*, 483–503.

Axtell, R. L. (2001). Zipf distribution of U.S. firm sizes. *Science, 293*, 1818–1820.

Azariadis, C., & Stachurski, J. J. (2006). Poverty traps. In P. Aghion & S. Durlauf (Eds.), *Handbook of Economic Growth*. Amsterdam: Elsevier.

Bagnoli, M., & Bergstrom, T. (2005). Log-concave probability and its applications. *Economic theory, 26*, 445–469.

Bakirov, N. K. (1998). Nonhomogenous samples in the Behrens-Fisher problem. *Journal of Mathematical Sciences, 89*, 1460–1467.

Bakirov, N. K., & Székely, G. J. (2005). Student's *t*-test for Gaussian scale mixtures. *Zapiski Nauchnyh Seminarov POMI, 328*, 5–19.

© Springer International Publishing Switzerland 2015

M. Ibragimov et al., *Heavy-Tailed Distributions and Robustness in Economics and Finance*, Lecture Notes in Statistics 214, DOI 10.1007/978-3-319-16877-7

Bakos, Y., & Brynjolfsson, E. (1999). Bundling information goods: Pricing, profits and efficiency. *Management Science, 45,* 1613–1630.

Bakos, Y., & Brynjolfsson, E. (2000a). Aggregation and disaggregation of information goods: Implications for bundling, site licensing and micropayment systems. In D. Hurley, B. Kahin & H. Varian (Eds.), *Proceedings of Internet Publishing and Beyond: The Economics of Digital Information and Intellectual Property.* Cambridge, MA: MIT.

Bakos, Y., & Brynjolfsson, E. (2000b). Bundling and competition on the internet. *Marketing Science, 19,* 63–82.

Bakun, W. H., Johnston, A. C., & Hopper, M. G. (2003). Estimating locations and magnitudes of earthquakes in eastern North American from Modified Mercalli intensities. *Bulletin of the Seismological Society of America, 93,* 190–202.

Barro, R. J., & Sala-i-Martin, X. (2004). *Economic growth.* Cambridge: MIT.

Beirlant, J., Goegebeur, Y., Teugels, J., & Segers, J. (2004). *Statistics of extremes.* Wiley Series in Probability and Statistics. Chichester: Wiley; Theory and applications, With contributions from Daniel De Waal and Chris Ferro.

Berger, J. O. (1985). *Statistical decision theory and Bayesian analysis.* New York: Springer.

Bernanke, B. S. (1983). Nonmonetary effects of the financial crisis in the propagation of the great depression. *American Economic Review, 73,* 257–276.

Bickel, P. J., & Lehmann, E. L. (1975a). Descriptive statistics for nonparametric models. I. Introduction. *Annals of Statistics, 3,* 1038–1044.

Bickel, P. J., & Lehmann, E. L. (1975b). Descriptive statistics for nonparametric models. II. Location. *Annals of Statistics, 3,* 1045–1069.

Bosker, M., Brakman, S., Garretsen, H., de Jong, H., & Schramm, M. (2007). The development of cities in Italy 1300–1861. CESifo Working Paper No. 1893.

Bosker, M., Brakman, S., Garretsen, H., & Schramm, M. (2008). A century of shocks: The evolution of the German city size distribution 1925–1999. *Regional Science and Urban Economics, 38,* 330–347.

Bouchaud, J.-P. and Potters, M. (2004). *Theory of financial risk and derivative pricing: From statistical physics to risk management* (2nd edn.). Cambridge: Cambridge University Press.

Box, G. E. P., & Tiao, G. C. (1973). *Bayesian inference in statistical analysis.* Reading/London/Don Mills: Addison-Wesley.

Brillinger (1973). Estimation of the mean of a stationary time series by sampling. *Journal of Applied Probability 10,* 419–431.

Brumelle, S. L. (1974). When does diversification between two investments pay. *Journal of Financial and Quantitative Finance, 9,* 473–483.

Brunnermeier, M. (2009). Deciphering the liquidity and credit crunch 2007–2008. *Journal of Economic Perspectives, 23,* 77–100.

Brunnermeier, M., & Pedersen, L. H. (2009). Market liquidity and funding liquidity. *Review of Financial Studies, 22,* 2201–2238.

Caballero, R., & Krishnamurthy, A. (2008). Collective risk management in a flight to quality episode. *Journal of Finance, 63,* 2195–2230.

Campbell, J. Y., Lo, A. W., & MacKinlay, A. C. (1997). *The econometrics of financial markets*(2nd edn.). Princeton: Princeton University Press.

Candelon, B., & Straetmans, S. (2006). Testing for multiple regimes in the tail behavior of emerging currency returns. *Journal of International Money and Finance 25,* 1187–1205.

Carlin, B. P., & Louis, T. A. (2000). *Bayes and empirical Bayes methods for data analysis.* London: Chapman & Hall.

Carroll, S., LaTourrette, T., Chow, B. G., Jones, G. S., & Martin, C. (2005). *Distribution of losses from large terrorist attacks under the Terrorism Risk Insurance Act.* Santa Monica, Calif: Rand Corporation.

Chakraborty, I. (1999). Bundling decisions for selling multiple objects. *Economic Theory, 13,* 723–733.

Chan, W., Park, D. H., & Proschan, F. (1989). Peakedness of weighted averages of jointly distributed random variables. In L. J. Gleser, M. D. Perlman, S. J. Press & A. R. Sampson (Eds.), *Contributions to Probability and Statistics* (pp. 58–62). New York: Springer.

Chollete, L., de la Peña, V., & Lu, C.-C. (2012). International diversification: An extreme value approach. *Journal Of Banking & Finance, 36*, 871–885.

Christoffersen, P. F. (2012). *Elements of financial risk management* (2nd edn.). New York: Elsevier.

Chu, C. S., Leslie, P., & Sorensen, A. (2011). Bundle-size pricing as an approximation to mixed bundling. *American Economic Review, 101*, 263–303.

Conley, T. G. (1999). GMM estimation with cross sectional dependence. *Journal of Econometrics, 92*, 1–45.

Cont, R. (2001). Empirical properties of asset returns: Stylized facts and statistical issues. *Quantitative Finance, 1*, 223–236.

Cowell, F. A., & Flachaire, E. (2007). Income distribution and inequality measurement: The problem of extreme values. *Journal of Econometrics, 141*, 1044–1072.

Cummins, J. D. (2006). Should the government provide insurance for catastrophes? *Federal Reserve Bank of St. Louis Review, 88*, 337–379.

Cummins, J. D., Doherty, N., & Lo, A. (2002). Can insurers pay for the 'big one'? Measuring the capacity of the insurance market to respond to catastrophic losses. *Journal of Banking and Finance, 26*, 557–583.

Cummins, J. D., & Mahul, O. (2003). Optimal insurance with divergent beliefs about insurer total default risk. *The Journal of Risk and Uncertainty, 27*, 121–138.

Dansby, R. E., & Conrad, C. (1984). Commodity bundling. *American Economic Review, 74*, 377–381.

Davidson, R., & Flachaire, E. (2007). Asymptotic and bootstrap inference for inequality and poverty measures. *Journal of Econometrics, 141*, 141–166.

Davis, R., & Resnick, S. (1985). Limit theory for moving averages of random variables with regularly varying tail probabilities. *Annals of Probability, 13*, 179–195.

Davis, R. A., & Mikosch, T. (1998). The sample autocorrelations of heavy-tailed processes with applications to ARCH. *Annals of Statistics, 26*, 2049–2080.

de la Peña, V. H., & Giné, E. (1999). Decoupling: From dependence to independence. *Probability and Its Applications*. New York: Springer.

de la Peña, V. H., Ibragimov, R., & Sharakhmetov, S. (2003). On extremal distributions and sharp L_p-bounds for sums of multilinear forms. *Annals of Probability 31*, 630–675.

di Giovanni, J., Levchenko, A. A., & Ranciere, R. (2011). Power laws in firm size and openness to trade: Measurement and implications. *Journal Of International Economics 85*, 42–52.

Diamond, D., & Dybvig, P. (1983). Bank runs, deposit insurance, and liquidity. *Journal of Political Economy, 91*, 401–419.

Doukhan, P. (1994). *Mixing*. Lecture Notes in Statistics (Vol. 85). New York: Springer; Properties and examples.

Embrechts, P., Klüppelberg, C., & Mikosch, T. (1997). *Modelling Extremal Events for Insurance and Finance*. New York: Springer.

Embrechts, P., McNeil, A., & Straumann, D. (2002). Correlation and dependence in risk management: Properties and pitfalls. In M. A. H. Dempster (Ed.), *Risk Management: Value at Risk and Beyond* (pp. 176–223). Cambridge: Cambridge University Press.

Embrechts, P., Neslehova, J., & Wüthrich, M. (2009). Additivity properties for Value-at-Risk under Archimedean dependence and heavy-tailedness. *Insurance: Mathematics and Economics, 44*, 164–169.

Fama, E. (1965a). The behavior of stock market prices. *Journal of Business, 38*, 34–105.

Fama, E. (1965b). Portfolio analysis in a stable Paretian market. *Management Science, 11*, 404–419.

Fama, E. F., & MacBeth, J. (1973). Risk, return and equilibrium: Empirical tests. *Journal of Political Economy, 81*, 607–636.

Fang, H., & Norman, P. (2006). To bundle or not to bundle. *Marketing Science, 37*, 946–963.

Flannery, M. (2005). No pain, no gain: Effecting market discipline via reverse convertible debentures. In H. Scott (Ed.), *Capital Adequacy beyond Basel: Banking, Securities, and Insurance* (Chap. 5, pp. 171–197.). Oxford: Oxford University Press.

Fofack, H., & Nolan, J. P. (2001). Distribution of parallel exchange rates in African countries. *Journal of International Money and Finance, 20*, 987–1001.

Fölmer, H., & Schied, A. (2002). Convex measures of risk and trading constraints. *Finance and Stochastics, 6*, 429–447.

Frittelli, M., & Gianin, E. R. (2002). Putting order in risk measures. *Journal of Banking & Finance, 26*, 1473–1486.

Froot, K. A. (2001). The market for catastrophe risk: A clinical examination. *Journal of Financial Economics, 60*, 529–571.

Froot, K. A., & Posner, S. (2002). The pricing of event risks with parameter uncertainty. *The Geneva Papers on Risk and Insurance, 27*(2), 153–165.

Froot, K. A., Scharfstein, D., & Stein, J. (1993). Risk management: Coordinating corporate investment and financing policies. *Journal of Finance, 48*, 1629–1658.

Froot, K. A., & Stein, J. (1998). Risk management, capital budgeting, and capital structure policy for financial institutions: An integrated approach. *Journal of Financial Economics, 47*, 55–82.

Gabaix, X. (1999). Zipf's law and the growth of cities. *American Economic Review, 89*, 129–132.

Gabaix, X. (2009). Power laws in economics and finance. *Annual Review of Economics, 1*, 255–293.

Gabaix, X., Gopikrishnan, P., Plerou, V., & Stanley, H. E. (2006). Institutional investors and stock market volatility. *Quarterly Journal of Economics, 121*, 461–504.

Gabaix, X., & Ibragimov, R. (2011). Rank$-1/2$: A simple way to improve the OLS estimation of tail exponents. *Journal of Business and Economic Statistics, 29*, 24–39. Supplementary material: http://pages.stern.nyu.edu/~xgabaix/papers/AdditionalResults.pdf.

Gabaix, X., & Landier, A. (2008). Why has CEO pay increased so much? *Quarterly Journal of Economics 123*, 49–100.

Gordy, M. B. (2000). Credit VaR and risk-bucket capital rules: A reconciliation. *Federal Reserve Bank of Chicago (May)* (pp. 406–417).

Gordy, M. B. (2003). A risk-factor based model foundation for ratings-based bank capital rules. *Journal of Financial Intermediation, 12*, 199–232.

Granger, C. W. J., & Orr, D. (1972). Infinite variance and research strategy in time series analysis. *Journal of the American Statistical Association, 67*, 275–285.

Hanks, T., & Johnston, A. (1992). Common features of the excitation and propagation of strong ground motion for north american earthquakes. *Bulletin of the Seismological Society of America, 82*, 1–23.

Hanks, T., & Kanamori, H. (1979). A moment magnitude scale. *Journal of Geophysical Research, 84*, 2348–2350.

Hansen, B. E. (2015). The integrated mean squared error of series regression and a Rosenthal hilbert-space inequality. *Econometric Theory, 31*, 337–361.

Helpman, E., Melitz, M. J., & Yeaple, S. R. (2004). Export versus FDI with heterogeneous firms. *American Economic Review, 94*, 300–316.

Hennessy, D. A., & Lapan, H. E. (2009). Harmonic symmetries of imperfect competition on circular city. *Journal of Mathematical Economics, 45*, 124–146.

Hinloopen, J., & van Marrewijk, C. (2012). Power laws and comparative advantage. *Applied Economics, 44*, 1483–1507.

Hitt, L. M., & Chen, P.-y. (2005). Bundling with customer self-selection: A simple approach to bundling low-marginal-cost goods. *Management Science, 51*, 1481–1493.

Ho, P., Sussman, J., & Veneziano, D. (2001). Estimating the direct and indirect losses from a Midwest earthquake. Report SE-10, Mid-America Earthquake Center.

Hogg, R., & Klugman, S. (1983). On the estimation of long tailed skewed distributions with actuarial applications. *Journal of Econometrics, 23*, 91–102.

Horn, R. A., & Johnson, C. R. (1990) *Matrix Analysis*. Cambridge: Cambridge University Press; Corrected reprint of the 1985 original.

Hsieh, P. (1999). Robustness of tail index estimation. *Journal of Computational Graphical Statistics, 8*, 318–332.

Hubbard, G., Deal, B., & Hess, P. (2005). The economic effect of federal participation in terrorism risk. *Risk Management & Insurance Review, 8*, 177–303.

Huisman, R., Koedijk, K. G., Kool, C. J. M., & Palm, F. (2001). Tail-index estimates in small samples. *Journal of Business and Economic Statistics, 19*, 208–216.

Ibragimov, M., & Ibragimov, R. (2008). Optimal constants in the rosenthal in- equality for random variables with zero odd moments. *Statistics and Probability Letters, 78*, 186–189.

Ibragimov, M., & Ibragimov, R. (2014a). Heavy tails and upper tail inequality. Working paper, Kazan (Volga Region) Federal University and Imperial College Business School.

Ibragimov, M., & Ibragimov, R. (2014b). Unemployment and output dynamics in CIS countries: Okun's law revisited. Working paper, Kazan (Volga Region) Federal University and Imperial College Business School.

Ibragimov, M., Ibragimov, R., & Kattuman, P. (2013). Emerging markets and heavy tails. *Journal of Banking and Finance, 37*, 2546–2559.

Ibragimov, R. (2005). New majorization theory in economics and martingale convergence results in econometrics. Ph.D. Dissertation, Yale University.

Ibragimov, R. (2007). Efficiency of linear estimators under heavy-tailedness: Convolutions of α-symmetric distributions. *Econometric Theory, 23*, 501–517.

Ibragimov, R. (2009a). Heavy-tailed densities, In S. N. Durlauf & L. E. Blume (Eds.), *The New Palgrave Dictionary of Economics Online*. Palgrave Macmillan. http://www.dictionaryofeconomics.com/article?id=pde2008_H000191.

Ibragimov, R. (2009b). Portfolio diversification and value at risk under thick-tailedness. *Quantitative Finance, 9*, 565–580.

Ibragimov, R. (2014). On the robustness of location estimators in models of firm growth under heavy-tailedness. *Journal of Econometrics* . In press, http://dx.doi.org/10.1016/j.jeconom.2014.02.005. Also available as Harvard Institute of Economic Research Discussion Paper No. 2087. http://papers.ssrn.com/sol3/papers.cfm?abstract_id=774865.

Ibragimov, R., Jaffee, D., & Walden, J. (2009). Nondiversification traps in catastrophe insurance markets. *Review of Financial Studies, 22*, 959–993.

Ibragimov, R., Jaffee, D., & Walden, J. (2010). Pricing and capital allocation for multiline insurance firms. *Journal of Risk and Insurance, 77*, 551–578.

Ibragimov, R., Jaffee, D., & Walden, J. (2011). Diversification disasters. *Journal of Financial Economics, 99*, 333–348.

Ibragimov, R., Jaffee, D., & Walden, J. (2012). Insurance equilibrium with monoline and multiline insurers. Working paper, Imperial College Business School and the University of California at Berkeley.

Ibragimov, R., Mo, J., & Prokhorov, A. (2014). Fat tails and copulas: Limits of diversification revisited. Working paper, Imperial College Business School and the University of Sydney Business School.

Ibragimov, R., & Müller, U. K. (2010). *t*-statistic based correlation and heterogeneity robust inference. *Journal of Business and Economic Statistics, 28*, 453–468. Supplementary material: *https://www.princeton.edu/~umueller/tstat_sup.pdf*.

Ibragimov, R., & Müller, U. K. (2013). Inference with few heterogeneous clusters. Review of Economics and Statistics (forthcoming).

Ibragimov, R., & Sharakhmetov, S. (1997). On an exact constant for the Rosenthal inequality. *Theory of Probability and Its Applications, 42*, 294–302.

Ibragimov, R., & Sharakhmetov, S. (2002). Bounds on moments of symmetric statistics. *Studia Scientiarum Mathematicarum Hungarica, 39*, 251–275 (Date of submission: 1996).

Ibragimov, R., & Walden, J. (2007). The limits of diversification when losses may be large. *Journal of Banking and Finance, 31*, 2551–2569.

Ibragimov, R., & Walden, J. (2010). Optimal bundling strategies under heavy-tailed valuations. *Management Science, 56*, 1963–1976.

Ioannides, Y. M., Overman, H. G., Rossi-Hansberg, E., & Schmidheiny, K. (2008). The effect of information and communication technologies on urban structure. *Economic Policy, 54*, 201–242.

Jaffee, D. (2006a). Commentary on Should the government provide insurance for catastrophes? *Federal Reserve Bank of St. Louis Review, 88*, 381–385.

Jaffee, D. (2006b). Monoline restrictions, with applications to mortgage insurance and title insurance. *Review of Industrial Organization, 28*, 88–108.

Jaffee, D. (2009). Monoline regulations to control the systemic risk created by investment banks and GSEs. *The B.E. Journal of Economic Analysis & Policy, 9*(3). Available at http://www.bepress.com/bejeap/vol9/iss3/art17.

Jaffee, D., & Russell, T. (2006). Should governments provide catastrophe insurance? *The Economists' Voice, 3*(5). Available at http://www.bepress.com/ev/vol3/iss5/art6.

Jensen, D. R. (1997). Peakedness of linear forms in ensembles and mixtures. *Statistics and Probability Letters 35*, 277–282.

Jovanovic, B., & Rob, R. R. (1987). Demand-driven innovation and spatial competition over time. *Review of Economic Studies, 54*, 63–72.

Jurečková, J., & Sen, P. K. (1996). *Robust statistical procedures: Asymptotics and interrelations*, New York: Wiley.

Kagan, Y., & Knopoff, L. (1984). A stochastic model of earthquake occurrence. In *Proceedings of the 8th International Conference on Earthquake Engineering* (Vol. 1, pp. 295–302). Englewood Cliffs, NJ: Prentice-Hall.

Kamien, M., & Schwartz, N. (1982). *Market structure and innovation*. New York: Cambridge University Press.

Karlin, S. (1968). *Total positivity* (Vol. I). Stanford, CA: Stanford University Press.

Kashyap, A., Rajan, R., & Stein, J. (2008). Rethinking capital regulation. In *Federal Reserve Bank of Kansas City Symposium on Maintaining Stability in a Changing Financial System* (pp. 431–471).

Knopoff, L., & Kagan, Y. (1977). Analysis of the theory of extremes as applied to earthquake problems. *Journal of Geophysical Research, 82*, 5647–5657.

Kodes, L., & Pritsker, M. (2002). A rational expectations model of financial contagion. *Journal of Finance, 57*, 769–799.

Koedijk, K. G., Stork, P. A., & de Vries, C. G. (1992). Differences between foreign exchange rate regimes: The view from the tails. *Journal of International Money and Finance 11*, 462–473.

Kokoszka, P., & Wolf, M. (2004). Subsampling the mean of heavy-tailed dependent observations. *Journal of Time Series Analysis, 25*, 217–234.

Krishna, V. (2002). *Auction theory*. San Diego, CA: Academic.

Kunreuther, H., & Michel-Kerjan, E. (2006). Looking beyond TRIA: A clinical examination of potential terrorism loss sharing. *NBER Working paper No. 12069*. http://www.nber.org/papers/w12069.

Kyle, A., & Xiong, W. (2001). Contagion as a wealth effect. *Journal of Finance, 56*, 1401–1440.

Le Gallo, J., & Chasco, C. (2008). Spatial analysis of urban growth in Spain, 1900–2001. *Empirical Economics, 34*, 59–80.

Levy, M. (2003). Are rich people smarter? *Journal of Economic Theory, 110*, 42–64.

Levy, M., & Levy, H. (2003). Investment talent and the Pareto wealth distribution: Theoretical and experimental analysis. *Review Of Economics And Statistics, 85*, 709–725.

Lewbel, A. (1985). Bundling of substitutes or complements. *International Journal of Industrial Organization, 3*, 101–107.

Logan, B. F., Mallows, C. L., Rice, S. O., & Shepp, L. A. (1973). Limit distributions of self-normalized sums. *Annals of Probability, 1*, 788–809.

Loretan, M., & Phillips, P. C. B. (1994a). Testing the covariance stationarity of heavy-tailed time series. *Journal of Empirical Finance, 1*, 211–48.

Loretan, M., & Phillips, P. C. B. (1994b). Testing the covariance stationarity of heavy-tailed time series. *Journal of Empirical Finance, 1*, 211–248.

Ma, C. (1998). On peakedness of distributions of convex combinations. *Journal of Statistical Planning and Inference, 70*, 51–56.

Mahul, O., & Wright, B. (2004). Efficient risk sharing within a catastrophe insurance pool. Paper presented at the 2003 NBER Insurance Project Workshop. Available at http://www.arec.umd.edu/Department/Seminars/mahulwright0204.pdf.

Mandelbrot, B. (1963). The variation of certain speculative prices. *Journal of Business, 36*, 394–419.

Mandelbrot, B. (1997). *Fractals and scaling in finance. Discontinuity, concentration, risk*. New York: Springer.

Marshall, A. W., & Olkin, I. (1979). *Inequalities: Theory of majorization and its applications*. New York: Academic.

Marshall, A. W., Olkin, I., & Barro, A. (2011). *Inequalities: Theory of majorization and its applications* (2nd edn.). New York: Springer.

McAfee, R., M. J., & Whinston, M. (1989). Multiproduct monopoly, commodity bundling and correlation of values. *Quarterly Journal of Economics, 114*, 371–384.

McElroy, T., & Politis, D. N. (2002) Robust inference for the mean in the presence of serial correlation and heavy-tailed distributions. *Econometric Theory, 18*, 1019–1039.

McGuire, R. (2004). *Seismic Hazard and Risk Analysis*. Oakland, CA: Earthquake Engineering Research Institute.

McNeil, A. J., Frey, R., & Embrechts, P. (2005). *Quantitative risk management: Concepts, techniques, and tools*. Princeton: Princeton University Press.

Mikosch, T., & Stărică, C. (2000). Limit theory for the sample autocorrelations and extremes of a GARCH $(1, 1)$ process. *Annals of Statistics, 28*, 1427–1451.

Milgrom, P., & Weber, R. (1982). A theory of auctions and competitive bidding. *Econometrica, 50*, 1089–1122.

Nelson, C. R., & Plosser, C. I. (1982). Trends and random walks in macroeconomic time series. *Journal of Monetary Economics, 10*, 139–162.

Nešlehova, J., Embrechts, P., & Chavez-Demoulin, V. (2006). Infinite mean models and the lda for operational risk. *Journal of Operational Risk, 1*, 3–25.

Nze, P. A., & Doukhan, P. (2004). Weak dependence: Models and applications to econometrics. *Econometric Theory, 20*, 995–1045.

OECD (2005a). Policy issues in insurance no. 08: Catastrophic risks and insurance. OECD Publishing, August.

OECD (2005b). Policy issues in insurance no. 09: Terrorism risk insurance in OECD countries. OECD Publishing, July.

Ozsoylev, H. N., & Walden, J. (2011) Asset pricing in large information networks. *Journal of Economic Theory, 46*, 2252–2280.

Palfrey, T. R. (1983). Bundling decisions by a multiproduct monopolist with incomplete information. *Econometrica 51*, 463–483.

Patnaik, I., Shah, A., Sethy, A., & Balasubramaniam, V. (2011). The exchange rate regime in Asia: From crisis to crisis. *International Review of Economics and Finance, 20*(1), 32–43.

Payaslioğlu, C. (2009). A tail index tour across foreign exchange rate regimes in Turkey. *Applied Economics, 41*, 381–397.

Phillips, P. C. B. (1990). Time series regression with a unit root and infinite-variance errors. *Econometric Theory, 6*, 44–62.

Phillips, P. C. B., & Hajivassiliou (1987). Bimodal *t*-ratios. Cowles Foundation Discussion Paper No. 842. Available at http://cowles.econ.yale.edu/P/cd/d08a/d0842.pdf.

Phillips, P. C. B., & Solo, V. (1992). Asymptotics for linear processes. *Annals of Statistics, 20*, 971–1001.

Politis, D. N., Romano, J. P., & Wolf, M. (1999). *Subsampling*. Springer Series in Statistics. New York: Springer.

Pozo, S., & Amuedo-Dorantes, C. (2003). Statistical distributions and the identification of currency crises. *Journal of International Money and Finance, 22*, 591–609.

Proschan, F. (1965). Peakedness of distributions of convex combinations. *Annals of Mathematical Statistics, 36*, 1703–1706.

Quintos, C. E., Fan, Z., & Phillips, P. (2001). Structural change in tail behavior and the Asian financial crisis. *Review of Economic Studies, 68*, 633–663.

Rachev, S. T., & Mittnik, S. (2000). *Stable Paretian Models in Finance*, New York: Wiley.

Rochet, J. C., & Tirole, J. (1996). Interbank lending and systemic risk. *Journal of Money, Credit, and Banking, 28*, 773–762.

Romano, J. P., & Wolf, M. (1999). Subsampling inference for the mean in the heavy-tailed case. *Metrika 50*, 55–69.

Ross, S. A. (1976). A note on a paradox in portfolio theory. Philadelphia: Mimeo, University of Pennsylvania.

Rothschild, M., & Stiglitz, J. (1970). Increasing risk: I. A definition. *Journal of Economic Theory, 2*, 225–243.

Rousseeuw, P. J., & Leroy, A. M. (1987). *Robust regression and outlier detection*. New York: Wiley.

Roy, A. D. (1952). Safety first and the holding of assets. *Econometrica, 20*, 431–449.

Salinger, M. A. (1995). A graphical analysis of bundling. *Journal of Business, 68*, 85–98.

Samuelson, P. A. (1967a). Efficient portfolio selection for Pareto-Lévy investments. *Journal of Financial and Quantitative Analysis, 2*, 107–122.

Samuelson, P. A. (1967b). General proof that diversification pays. *The Journal of Financial and Quantitative Analysis, 2*, 1–13.

Schmalensee, R. L. (1984). Gaussian demand and commodity bundling. *Journal of Business, 57*, S211–S230.

Schwartz, D., & Coppersmith, K. (1984). Fault behavior and characteristic earthquakes. *Journal of Geophysical Research, 89*, 5681–5698.

Shaffer, S. (1994). Pooling intensifies joint failure risk. *Research in Financial Services, 6*, 249–280.

Silverberg, G., & Verspagen, B. (2007). The size distribution of innovations revisited: An application of extreme value statistics to citation and value measures of patent significance. *Journal of Econometrics, 139*, 318–339.

Sornette, D., Knopoff, L., Kagan, Y., & Vanneste, C. (1996). Rank-ordering statistics of extreme events: Application to the distribution of large earthquakes. *Journal of Geophysical Research, 101*, 13883–13894.

Stock, J. H., & Watson, M. W. (2007). *Introduction to Econometrics* (2nd edn.). New York: Addison-Wesley.

Stremersch, S., & Tellis, G. J. (2002). Strategic bundling of products and prices: A new synthesis. *Journal of Marketing, 66*, 55–72.

SwissRe (2006). Sigma. *Sigma, 2*, 267–270.

Toda, A. A. (2012). The double power law in income distribution: Explanations and evidence. *Journal Of Economic Behavior & Organization, 84*, 364–381.

Tong, Y. L. (1994). Some recent developments on majorization inequalities in probability and statistics. *Linear Algebra and its Applications, 199*, 69–90.

Uchaikin, V. V., & Zolotarev, V. M. (1999). *Chance and stability. Stable distributions and their applications*. Utrecht: VSP.

Utev, S. A. (1985). Extremal problems in moment inequalities. *Proceedings of the Mathematical Institute of the Siberian Branch of the USSR Academy of Sciences, 5*, 56–75. (In Russian).

Venkatesh, R., & Kamakura, W. (2003) Optimal bundling and pricing under a monopoly: Contrasting complements and substitutes from independently valued products. *Journal of Business 76*, 211–231.

Volcker, P. A. (2010). Statement for the committee on banking, housing, and urban affairs of the United States Senate, February 2, 2010. Available at http://banking.senate.gov/public/index.cfm?FuseAction=Files.View&FileStore_id=ec787c56-dbd2-4498-bbbd-ddd23b58c1c4.

Wagner, W. (2010). Diversification at financial institutions and systemic crises. *Journal of Financial Intermediation, 19*, 373–386.

Woo, G. (1999). *The Mathematics of Natural Catastrophes*. London: Imperial College Press.

Wu, S.-y., Hitt, L. M., Chen, P.-y., & Anandalingam, G. (2008). Customized bundle pricing for information goods: A nonlinear mixed-integer programming approach. *Management Science, 54*, 608–622.

Zhang, J., Chen, Q., & Wang, Y. (2009). Zipf distribution in top Chinese firms and an economic explanation. *Physica A: Statistical mechanics and its explanation, 388*, 2020–2024.

Zolotarev, V. M. (1986). *One-dimensional stable distributions*. Providence: American Mathematical Society.